Cave and Karst Systems of the World

Series Editor

James W. LaMoreaux, P.E.LaMoreaux and Associates, Tuscaloosa, AL, USA

More information about this series at http://www.springer.com/series/11987

Eric Gilli

Port Miou and Le Bestouan (Cassis, France)

The Largest French Submarine Karst Springs

Springer

Eric Gilli
Department of Geography
University of Paris 8
Saint-Denis, France

ISSN 2364-4591 ISSN 2364-4605 (electronic)
Cave and Karst Systems of the World
ISBN 978-3-030-50191-4 ISBN 978-3-030-50192-1 (eBook)
https://doi.org/10.1007/978-3-030-50192-1

This Springer imprint is published by the registered company Springer Nature Switzerland AG
The registered company address is: Gewerbestrasse 11, 6330 Cham, Switzerland

Foreword

Port Miou is an extraordinary speleological, scientific and human adventure. In Southern France, close to the small city of Cassis, at the gates of Marseille City, a gigantic and mysterious brackish river flows under the arid landscape of the Calanques massif. Known from the Antiquity, this underground river could only be studied after the invention of the scuba diving in the 1950s. Seventy years later, although a tremendous amount of information was collected in the 1970s then from 2000, the mystery of its origin is still present.

Economic interest, attraction of the unknown, scientific questions have been the motor of a long series of grouped or individual actions, sometimes titanic, sometimes tragic, where explorers, scientists or managers, have associated or confronted each other around three main issues:

- How long are the caves, where do they extend?
- Where does Port Miou water come from?
- Why is the water brackish?

Port Miou is an area where the notion of mutual contribution finds all its grandeur. Countless are those who, famous or unknown, have brought their brick to the construction of the knowledge of this complex system. Technological progress, speleological exploration and scientific research are closely intertwined in that adventure.

I entered in Port Miou cave history in the late 1970s when I started becoming a caver. A member of our cavers group had told us that he had been diving there. He had explained us that the construction of a submarine dam was in progress into the cave. He had been allowed to join the divers who were in charge of the work. His description was very attractive, but the place was closed, and the access was forbidden thus visiting Port Miou was no more than a secret dream. The Port Miou project failed a few years later, and I did not heard about it for a long time.

However in the 1990s, I started to study several submarine springs around Nice (South-eastern France), close to the Italian border. I could inventory and quantify the water outlets that were flowing, either on the shore, or below the sea surface. The purpose was mainly to catch the water, as this area was suffering a lack of water during the summer. A second aim was to estimate the total amount of water that was flowing from the local karst units in view to equilibrate the hydrologic balance of that zone. The catchment failed but the collected data gave valuable information on several karst systems which made it possible to equilibrate the balance.

This study, that was supported by the French Ministry of Environment, was considered as a success, and a few months later, I was asked by the Regional Water Agency to inventory the coastal and submarine karst springs in the whole Southeastern France, including Corsica island. This was an opportunity to collect and study many documents that concerned Port Miou. I was surprised by the important discharge of the springs, measured by the technicians during the construction of the dam in the 1970s.

In addition, at that time, three important discoveries had changed my view on karst systems:

- During their explorations, cave divers had reached a depth of 174 m in Port Miou, in a shaft, far from the entrance;
- The model of a Mediterranean salinity crisis during the Messinian was widely accepted, and several authors had described their effects on coastal karsts;
- After the discovery of the prehistoric Cosquer Cave, a colleague of mine had realized a bathymetric map that was presenting a karst plateau, 150 m below the sea level.

Thanks to my experience on Nice submarine springs I quickly understood that Port Miou was a gigantic and old system that was related to the Messinian salinity crisis. This was the beginning of a new Port Miou adventure.

In this book, I describe the site, the previous works, and I present my Messinian model and the recent explorations that confirm it. The purpose of the book is also to share, with as many as possible, this passion for this phenomenon which will remain mysterious for a few decades and which will certainly promote new vocations.

Nice (France) Eric Gilli

Contents

1 Presentation of the Cassis Springs

1.1 Location

Both Cassis springs, Port Miou and Le Bestouan, flow at sea in the Calanques massif, a limestone area that extends east of Marseille, in Southern France (Fig. 1.1). They are located in the vicinity of Cassis, a small harbor (Fig. 1.2). It is a touristic place where tens of thousands persons come each year to visit the Calanques massif by boat or by foot and to enjoy the pleasures of sea. It is also a place famous for its vineyard (Fig. 1.3). The springs are the outlets of an important karst aquifer that discharges to the Mediterranean Sea (Fig. 1.4).

1.1.1 Le Bestouan

Le Bestouan is located at the foot of a limestone cliff, close to the entrance of Cassis harbor. The entrance is 1.5 m below the sea level (Fig. 1.5).

1.1.2 Port Miou

Port Miou is located west of Cassis, in a Calanque, the local name for a narrow creek (Figs. 1.6 and 1.7). This spring is known from immemorial time and was cited, 2600 years ago, by the Greek geographer Pytheas. It flows from a flooded gallery, 6 m below the sea level, at the foot of a limestone cliff. Several outlets are visible in the creek (Fig. 1.8), and the strong current of brackish water that escapes from the cave is clearly perceptible when the sea is calm (Figs. 1.9 and 1.10).

On the limestone shore, upstream of the spring, two natural shafts join the cave. The largest one is now occupied by discharge waste facilities from the bauxite treatment plant of Gardanne (Pechiney then Rio-Tinto-Alcan then Alteo). Since 1966, important quantity of red mud is dumped 350 m below the sea level in the Cassidaigne submarine canyon, a few kilometers off Cassis. The red mud's pipeline is placed in the shaft and into the last portion of the cave; then, it crosses Port Miou creek and heads on the sea floor until the canyon. This environmental damage is regularly denounced by environmental activists but the discharge continues.

1.2 Description of the Caves

1.2.1 Le Bestouan (from Douchet 2012)

It is a flooded cave with a succession of low points down to −29 m. It is about 4 km long with two main galleries (Figs. 1.11 and 1.12). Its exploration began at the same time as Port Miou spring, from the invention of the scuba diving equipment in the 1960s. The ultimate points were reached by diving in 1993 on a narrow passage and a scree, impassable without a clearing that is considered too dangerous given the distance. Close to that place a chamber was discovered, the Volanthen room, whose ceiling is 35 m height. There is no survey of the cave, but five radiolocations were made in different air bells.

Grande Gallerie (Great Gallery): The water flows from many outlets at the base of the cliff. Only two conduits are accessible. The main entrance is a narrow conduit 1 m in diameter, 1 m below the surface of the water, with a strong current, even at low water. This gallery remains shallow for 30 m and leads to the top of a 15-m-deep pitch. At its base the gallery is larger: 3 to 6 m wide, 3 to 4 m high. The depth varies between −20 and −24 m. The floor is covered with an important clay deposit. At point 400, the conduit is washed

E. Gilli, *Port Miou and Le Bestouan (Cassis, France)*, Cave and Karst Systems of the World,
https://doi.org/10.1007/978-3-030-50192-1_1

Fig. 1.1 Location of Cassis springs

Fig. 1.3 Cassis vineyard and Canaille Cape in the background

out by the current (Fig. 1.13). The depth reaches −27 m along a large flowstone deposit. Then, the gallery becomes narrow and the current is more violent. It is very muddy, and the visibility is poor.

At point 650, the river flows in a reduced section (1 × 5), the passage is over a clay dune. A rope makes it possible to face the strong current on a slope inclined 45°. On the left, there is a side passage. At point 700, there is an air bell, and then the gallery becomes a flattener that plunges quickly. The floor is sandy.

From point 700 m, the gallery is large, with a succession of wide passages and very large rooms. The depth varies between −22 and −28 m.

Fig. 1.2 Harbor of Cassis and Canaille Cape in the background

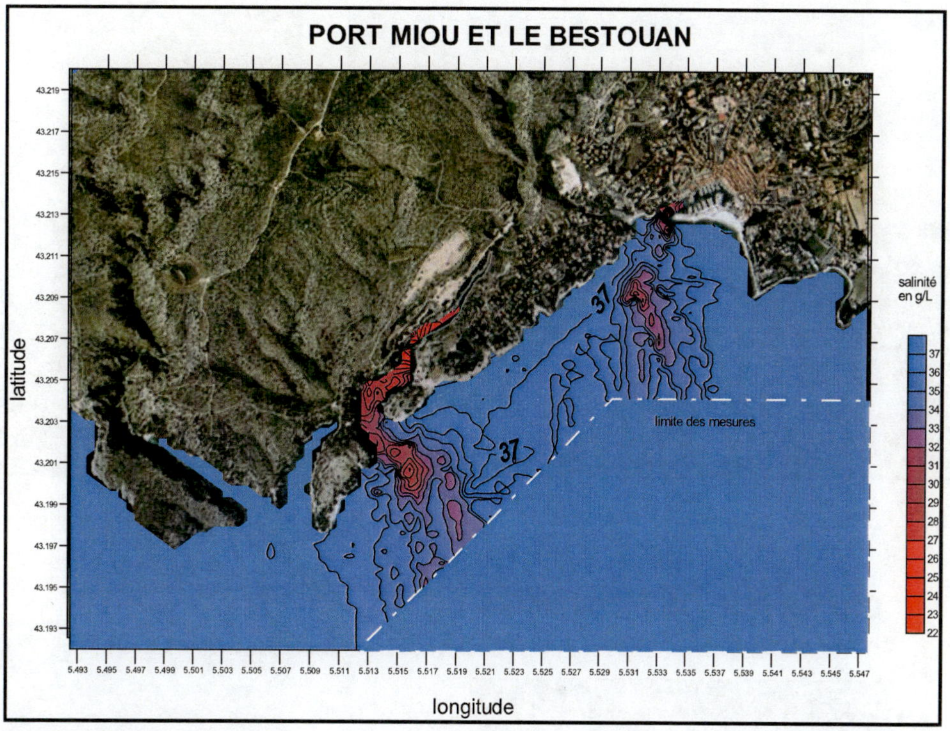

Fig. 1.4 Surface salinity map of Cassis area showing Port Miou and Le Bestouan plumes of brackish water (Gilli et al. 2009)

Fig. 1.5 Underwater view of Le Bestouan entrance. The rope makes it possible to face the strong current that flows from the cave (*Photo* L. Dendeloeuf)

Fig. 1.6 Entrance of Port Miou creek

At point 1400 m, a 20-m-long air bell forms a narrow place where it is impossible to swim at surface. Then, the passage is a shredded meander 1 m wide and 2–4 m high for around 100 m long. After that, the gallery becomes larger (3 × 3) the depth being around −15 m. The general direction fluctuates from north to northeast. Various passages, some of which are very chaotic, lead to a 5 × 5 room at 2900 m from the entrance. It is the junction of three galleries in which the current is very important, but they become impenetrable after about 20 m.

Galerie du Flou (Blur gallery): At 2440 m from the entrance at −25 m depth, a zone of turbulence (blur) hides the departure, on the left bank, of a large parallel gallery

Fig. 1.7 Calanque of Port Miou

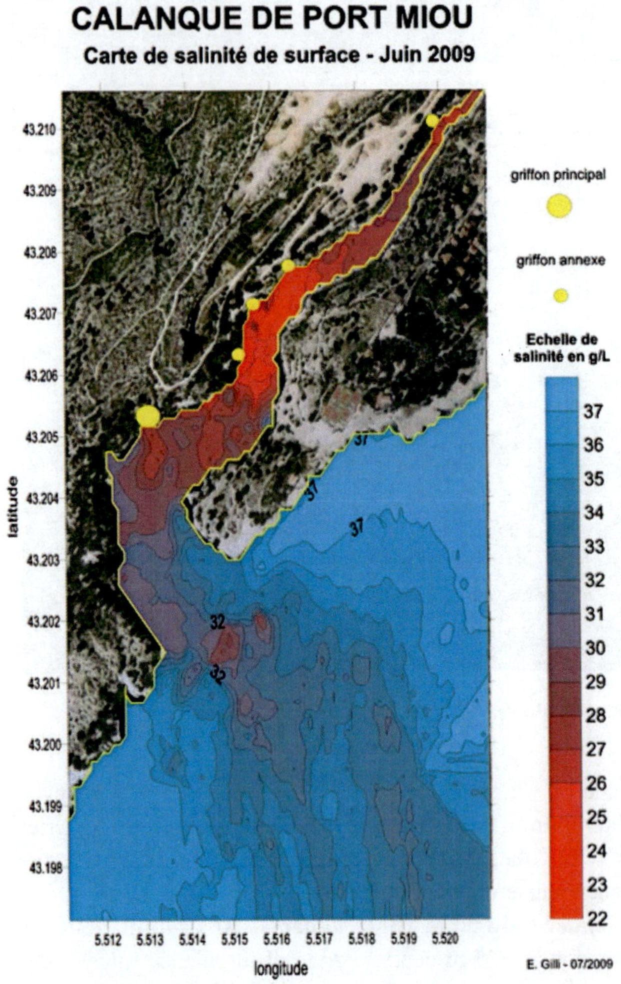

Fig. 1.8 Salinity map of Port Miou creek (Gilli et al. 2009)

Fig. 1.9 Brackish water current in Port Miou (visible between the two persons)

Fig. 1.10 Brackish water of Port Miou spring

where the current is as powerful as in the Great Gallery. After about 500 m the gallery stops on a breakdown zone where a narrow passage, between the boulders, gives access to a chamber, the Volanthen room, whose ceiling is 35 m high. It is blocked with a boulder slope. The water flows through the slope, emerging from gaps in the boulders. No way was found to go further.

The total development of flooded conduits explored in Le Bestouan is around 4 km.

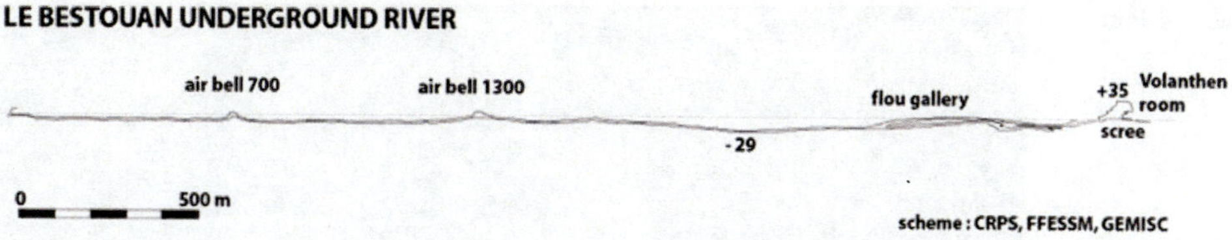

Fig. 1.11 Cross section of Le Bestouan cave

Fig. 1.12 Cross section of Le Bestouan cave (heights are exaggerated)

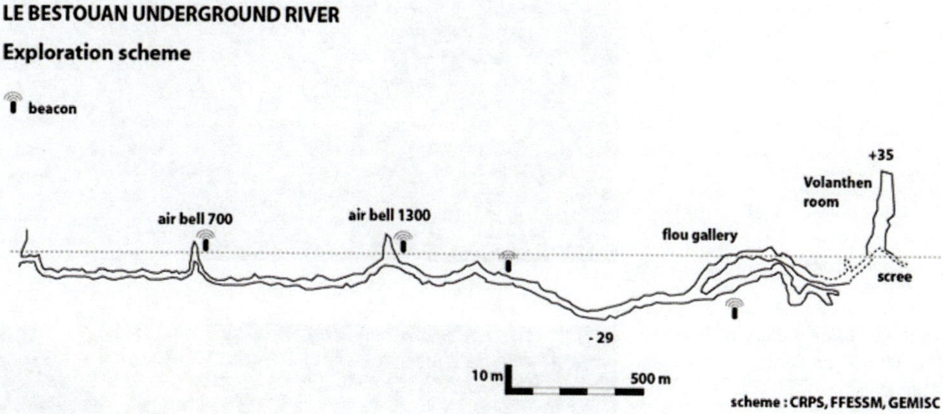

Fig. 1.13 Great gallery in Le Bestouan (*Photo* L. Dendeloeuf)

1.2.2 Port Miou

Description of the cave

Port Miou cave network is totally flooded. It is huge cave where galleries overpass 20 m in diameter (Figs. 1.14, 1.15, 1.16, 1.17 and 1.18). The cave is explored since the 1960s. The total length is more than 2500 m. The cave is divided in four parts.

From the entrance, a 530-m-long gallery leads to a dam that was built in an air bell during the 1970s. From that point, a large gallery, with some lateral passages and a maze on the right side, extends until another air bell located 2200 m from the entrance.

At the end of the main gallery, there is a very deep and huge shaft that was discovered in 1981 (Fig. 1.19) (Douchet 1993).

The distance from the entrance and the important depth has long been an almost insurmountable difficulty. However, in 2016, thanks to the use of closed-circuit rebreathers (CCR), the bottom of the pitch was reached and an ultimate point, in an horizontal gallery, was explored until a depth of 233 m. The karst conduit continues beyond, and exploration projects are still going on (Meniscus 2017).

A project of water catchment took place in the 1970s (Potié 1974). Both springs were monitored. Then a 50-m-deep well and an horizontal gallery were dug to reach the air bell 500 discovered by the divers (Figs. 1.20, 1.21 and 1.22). From that place, two successive underground dams were built in the flooded cave (Fig. 1.23, 1.24 and 1.25). The project failed in 1973, and the place was abandoned until 2000 when I started new research on the site. New monitoring equipment was installed that are now enhanced and used by Marseille University.

Fig. 1.14 Main gallery of Port Miou (*Photo* Frederic Swierczynski)

Fig. 1.15 Main gallery of Port Miou (*Photo* Frederic Swierczynski

Fig. 1.16 Main gallery with important clay filling (*Photo* Frederic Swierczynski)

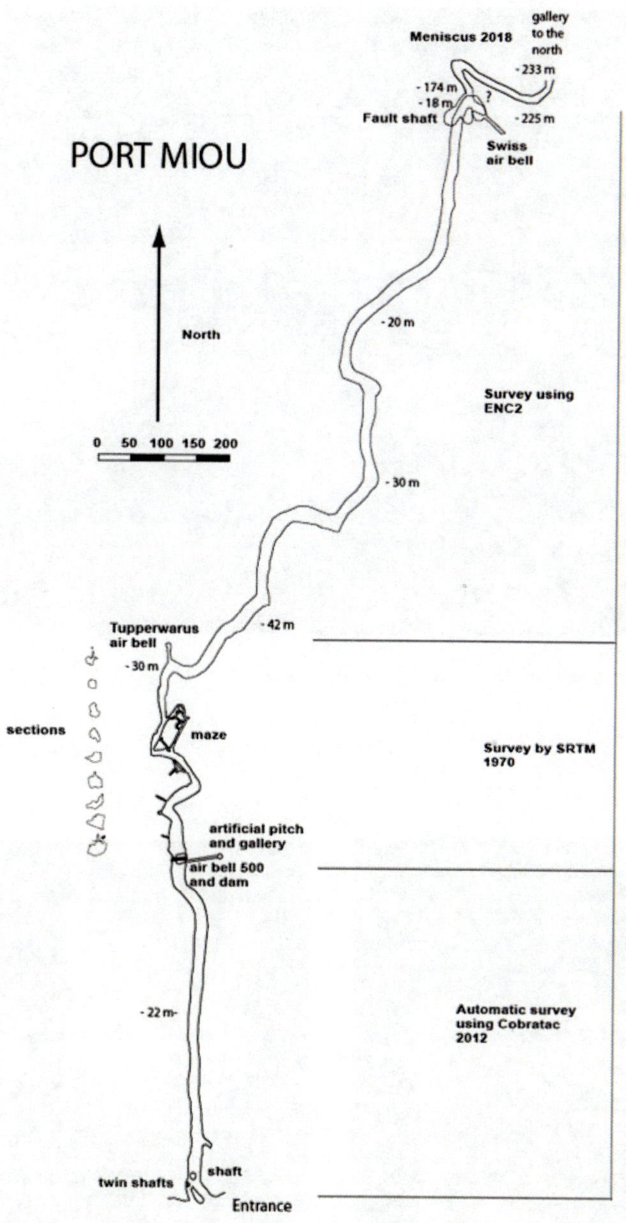

Fig. 1.17 Map of Port Miou cave

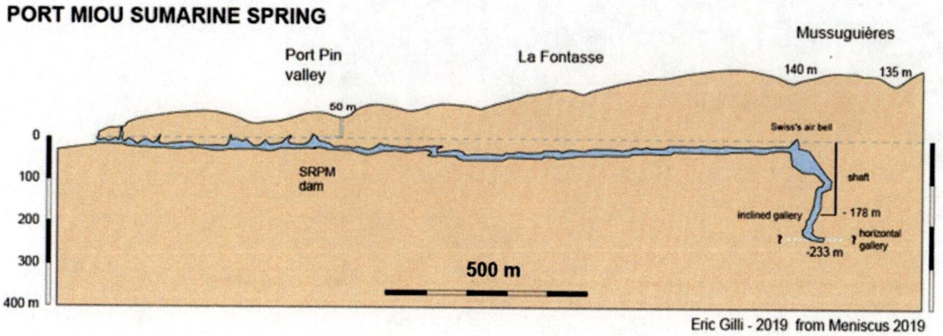

Fig. 1.18 Cross section of Port Miou cave

Fig. 1.19 Port Miou shaft with Stanton pendulum (*Photo* Swierczynski 2016)

Fig. 1.20 Pitch that makes it possible to reach the air bell 500

Fig. 1.21 Lift used until the 2010s

Fig. 1.22 Port Miou artificial gallery

Fig. 1.23 Underground and underwater dam in Port Miou (*Photo* Swierczynski 2016)

Fig. 1.24 Sump upstream of the dam (*Photo* Creative Common—JYB Devot)

Fig. 1.25 Sump downstream of the dam

References

Douchet, M.: La riviere sous-marine de Bestouan. A. N. A. R. Bull', **32**, 4–5 (2012)

Douchet, M.: Port Miou - Le Bestouan. Spelunca **49**, 5–6 (1993)

Gilli, E., Fournillon, A., Tassy, A., Arfib, B.: Localisation des émergences karstiques des calanques et de la baie de La Ciotat. Projet Karsteau – Université de Provence, p. 49 (2009)

Meniscus, X.: Encore plus profond à Port Miou: –233. Spelunca **146**, 27–29 (2017)

Potié, L.: Captage des résuregences sous-marines d'eau douce. Observations sur les effets du barrage expérimental de Port Miou, *BRGM, Marseille, 74 SGN 272*, 87, p. (1974)

Swierczynski, F.: Cave diving, Port Miou 2016 (U-W-X) https://vimeo.com/161296247 (2016)

2 Natural Context

2.1 Topography and Landscape

The Basse Provence is a limestone zone of low mountains that extends from the foothills of the Alps (Pre-Alps) in the northeast, to the sea in the south. It is limited, to the east, by the flat terranes of Rhône valley and to the west by a crystalline range, the Maures massif.

The landscape is of Mediterranean type, similar to Greece, Spain or Italy, with scrublands, pine trees and dry limestone areas where karst features are well developed.

The shore, between Marseille and Toulon, is formed by the Calanques massif. It is a high line of white limestone cliffs cut by a series of narrow creeks, locally called Calanques (Figs. 2.1, 2.2 and 2.3). From the west to the east, the main ones are: Callelongue, Sormiou, Morgiou, Sugiton, Le Devenson, L'Oule, En-Vau, Port Pin and Port Miou (Fig. 2.4).

A series of small islands, the Riou archipelago, is present in the western part of the shore. The Calanques massif ends at Cassis, where a depression in the topography is occupied by a small harbor city and its surrounding vineyards (Fig. 2.5). Then, to the west, the shore is formed again by a very high cliff, the Canaille Cape, that extends for several kilometers until another topographic trough, the Ciotat Bay.

To the north, the countryside is formed by a series of small mountains, the main ones being the Sainte Baume (altitude 1148 m) and the Sainte-Victoire in the north (altitude 1011 m). They surround valleys and karst depressions whose main ones are the Aubagne basin in the south, the Arc valley in the northwest and the Saint-Maximin plain to the northeast.

Four rivers cross the Basse Provence. In the north, the Arc flows to the west, until the Etang-de-Berre (Berre pond). In the northeast, the Argens, which is the main aerial river of this area, flows to the sea. In the southwest, the Huveaune circulates to the west, until the Marseille bay. In the southeast, the Gapeau flows to the sea, until the city of Hyères. These rivers are partially fed by karst springs.

The area contains two beautiful poljes in Cuges and Chibron, northeast of Cassis (Fig. 2.6).

2.2 Geology

2.2.1 Generalities

Basse Provence is classically divided into several parts that result from a complex history (Figs. 2.7 and 2.8 2.9).

- The Crystalline Provence;
- The Permian depression;
- The Limestone Provence and the Esterel;
- The Pre-Alps;
- The Valensole plateau;
- The Oligo-Miocene basins;
- The Rhodanian Provence.

The Cassis springs are located in the Limestone Provence.

2.2.2 Geological History

Paleozoic

During the Carboniferous (−360 to −295 Ma), the movement of the continents caused the uplift of the Hercynian range, which included the Pyrenees Provence chain where Provence, Corsica, Sardina formed a single block. The crystalline Provence that includes Maures, Esterel and Tanneron massifs is a remnant of this era.

E. Gilli, *Port Miou and Le Bestouan (Cassis, France)*, Cave and Karst Systems of the World,
https://doi.org/10.1007/978-3-030-50192-1_2

Fig. 2.1 Calanque massif

Fig. 2.2 Calanques landscape in Morgiou

Fig. 2.3 Sugiton Calanque and Torpilleur island

Fig. 2.6 Google Earth view of Cuges polje

Fig. 2.4 Scheme of the Calanques massif

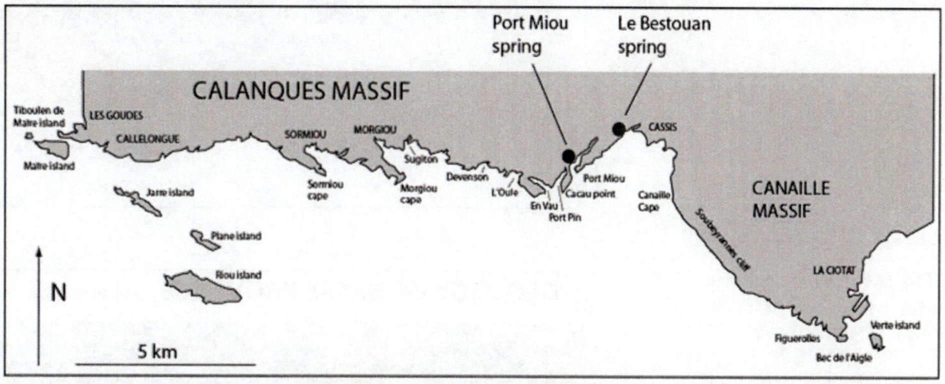

Fig. 2.5 Cassis from the top of the Canaille Cape cliffs

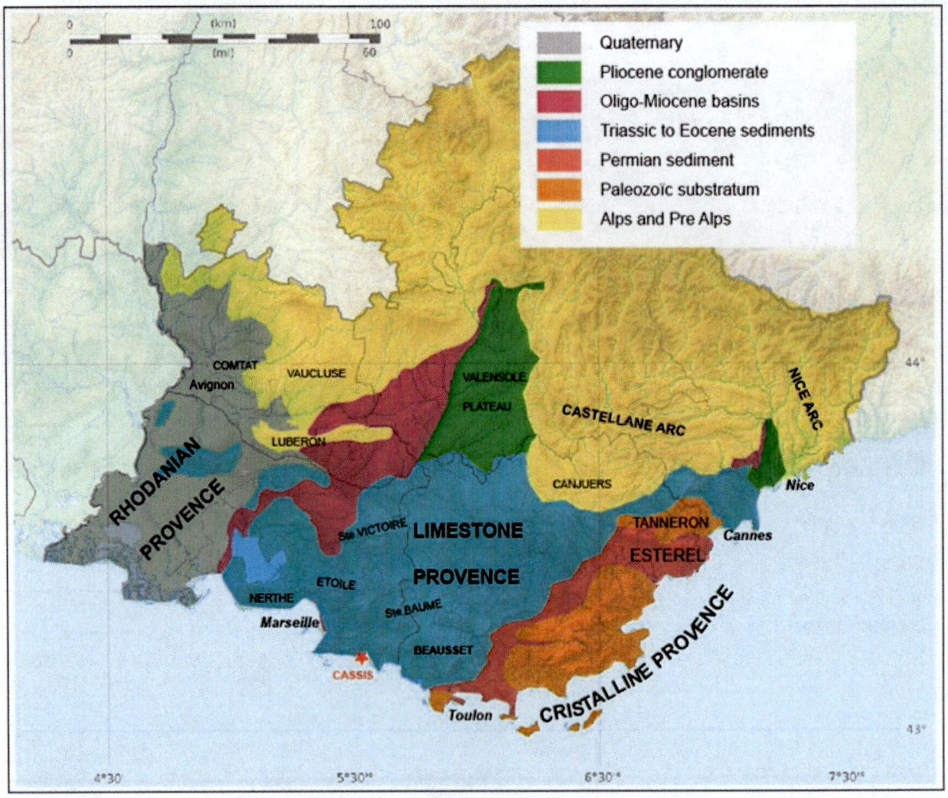

Fig. 2.7 Geology domains of Basse Provence

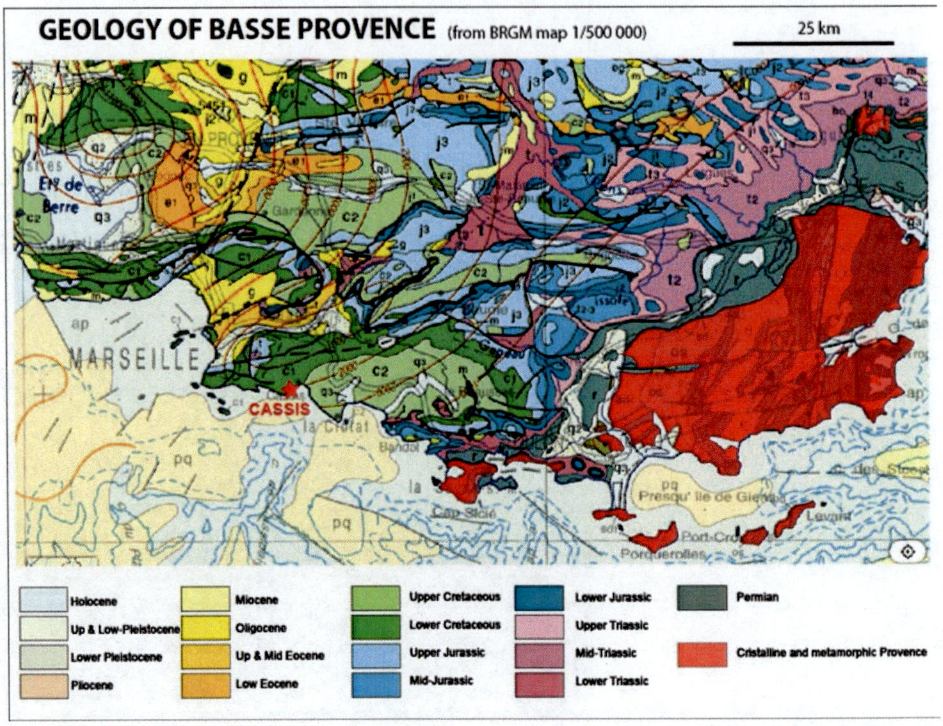

Fig. 2.8 Geological map of Basse Provence (from BRGM map 1/250,000)

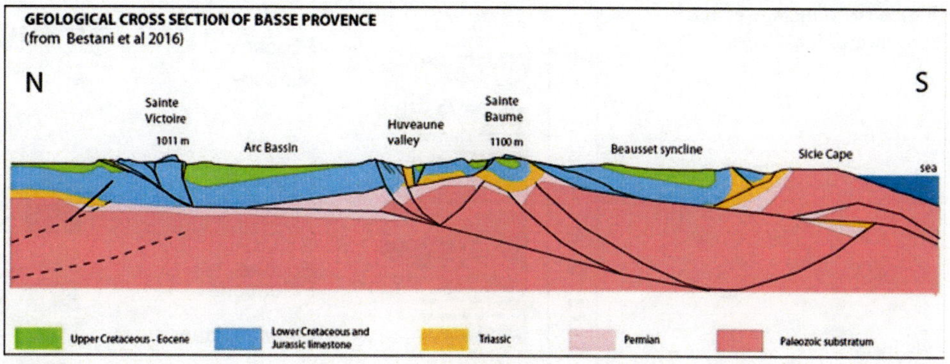

Fig. 2.9 Geological cross section of Basse Provence (from Bestani et al. 2016)

During the Permian, intense erosion provoked the deposit of a thick series of pink sediments. They are present in the depression that surrounds the Maures massif. An important volcanic activity took place around −275 Ma with important deposit of rhyolites that forms now the Esterel massif.

After the Hercynian uplift, the sea covered again the future Provence and deposited a thick series of carbonate sediments above the Hercynian metamorphic substratum and the Permian sediments.

Mesozoic (−245 to −65 Ma)

During the Triassic (−245 to 205 Ma), sandstone, dolomitic limestone, marls are deposited. At the end of the Triassic, a partial sea regression provoked the formation of thick layers of evaporites and clays that played an important role during the future tectonic phases.

During the Jurassic (−205 to −135 Ma) and the Lower Cretaceous (−135 to −96 Ma), a new marine transgression deposited varying thickness of carbonate sediment, marls, dolomite, Urgonian limestones or reef limestones.

Cenozoic (−65 to −1.8 Ma)

During the Eocene (−53 to −34 Ma), the Provence emerged.

In the Oligocene (−34 to −23.5 Ma), an intense erosion provoked the deposit of continental detrital sediments. The sliding movement of the Africa Plate changed, and the compression phase gave place to a distension one, during which the Rhône graben was formed west of Provence. This period also saw the opening of the Ligurian ocean, by rotation of the Corsica–Sardinia block that moved from Provence to its current position.

Numerous grabens formed in which marls, limestone and conglomerate sedimented.

In the Miocene (−23.5 to −5.3 Ma), the Africa and Eurasia plates converged. This started the uplift of the Alps, north of Provence and the movement toward the south of its sediments cover, which formed the folds and overthrusts of the Pre-Alps: Castellane arc to the northeast or Luberon to the north.

An important event occurred between 5.8 and 5.3 Ma: the Messinian salinity crisis. The tectonic movements of Africa Plate caused a separation of the Mediterranean Sea from the Atlantic which provoked a strong evaporation, the deposit of evaporites and the deepening of the rivers and karst aquifers. Then, the Strait of Gibraltar opened which restored the sea level in Mediterranean basin. The seawater filled up the deep valleys and formed important rias. A small sea basin was present north of Provence, in the Valensole area.

The rising up of the Alps provoked an uplift of the Provence and a tilting toward south. The Mediterranean Sea acquired its current position.

Quaternary (−1.8 Ma to now)

During the Quaternary, the different glacial periods provoked successive lowering of the Mediterranean Sea level down to 120 m.

2.2.3 Lithology

Due its size, Basse Provence offers important facies and thickness variations (Fig. 2.10).

Except for the crystalline part, in the southwestern part of Basse Provence, three main formations structure the landscape:

- Jurassic and Late Cretaceous limestones form the highest places in the landscape (Sainte-Baume, Sainte-Victoire, Calanques and Canaille cliffs …);
- Permian and Triassic sediments form large depressions;
- Early Cretaceous to Quaternary terranes form other low or perched depressions. But locally these formations, that also include hard rocks (sandstone, conglomerate, limestone) may form small mountains or cliffs.

A special attention has to be given to the mid-Cretaceous formations in the Beausset unit, east of Cassis (Aptian, Cenomanian and Turonian) which may or may not constitute a hydraulic barrier to groundwater circulations. This

Fig. 2.10 Lithology of Basse Provence (Tassy 2012)

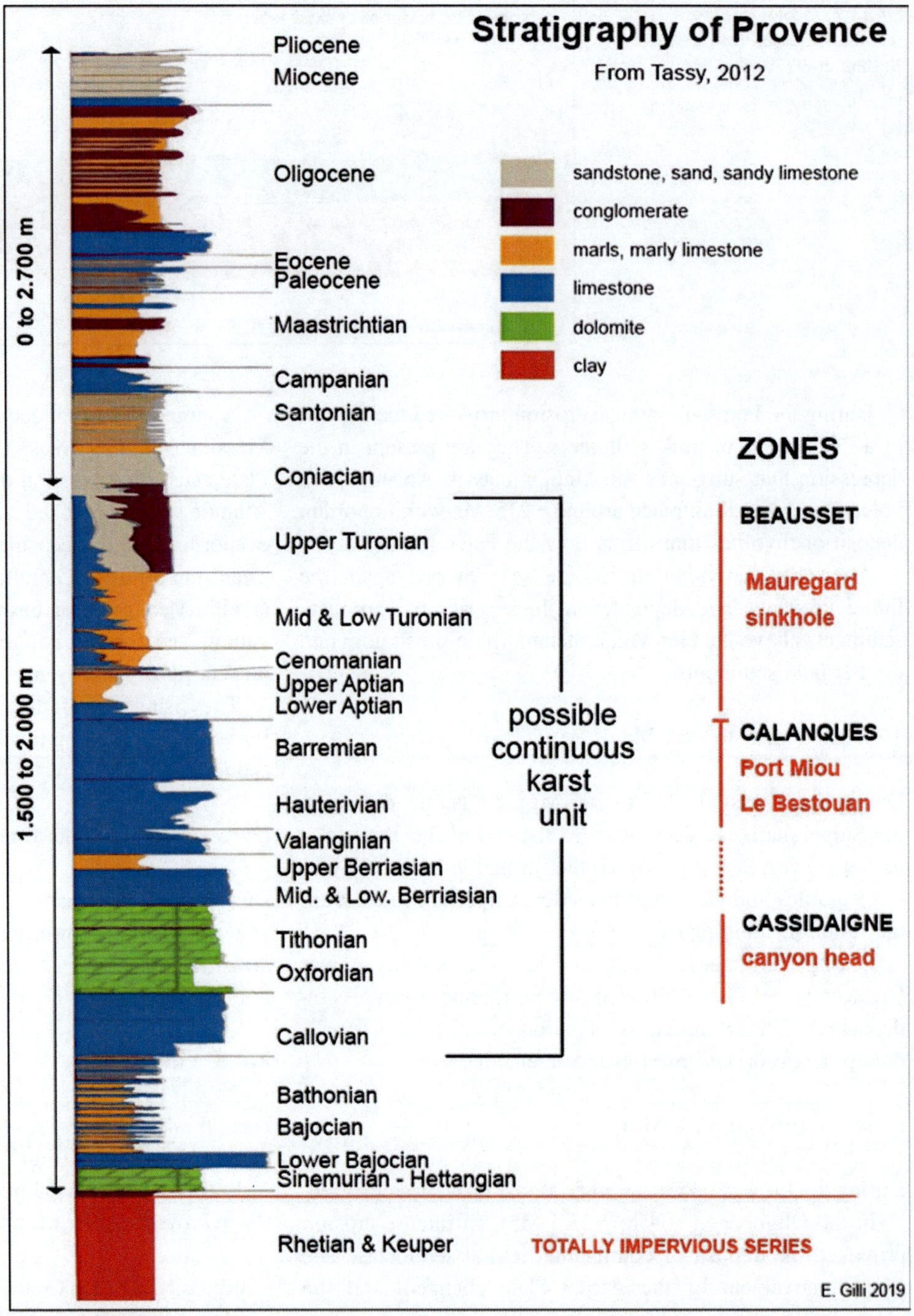

depends on their limestone or marly facies, but also on the presence of an important faulting. A similar observation is made for the mid-Jurassic (Bathonian and Callovian).

2.2.4 Tectonics and Geomorphology

As described previously, Provence experienced many tectonic phases, caused by the formation of the Pyrenees, the aperture of the Ligurian basin and the uplift of the Alps. Above the crystalline substratum, the sedimentary cover peeled off and slipped the north, then to the south. The main phases can be summarized as follows.

After the sedimentation of a thick series of sediments above the Hercynian substratum, the Provence emerged during the Albian (−108 to 96 Ma), mid-Cretaceous), and formed the "Durancian isthmus." Intense erosion on limestone took place and formed an important karst whose caves and depressions were filled with continental deposits. This material evolved later in bauxite.

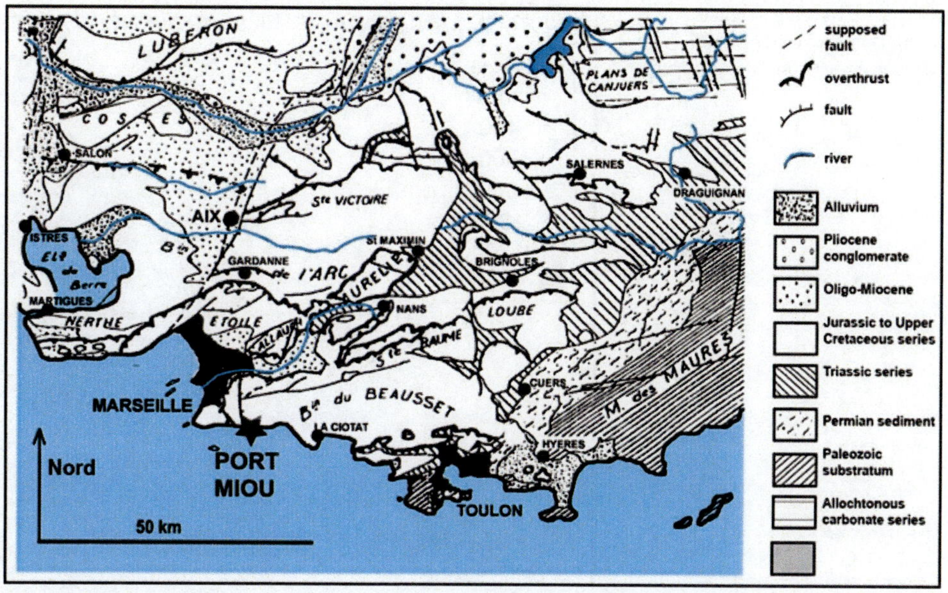

Fig. 2.11 Structure of Basse Provence (from **Gouvernet et al.** 1979)

At the end of the Cretaceous (−65 Ma), a new marine transgression sealed the karst; then, the Provence experienced a general uplift and a tilting movement from south to north.

The sliding of the Iberian plate toward the European plate provoked the uprising of the Pyrenees range. During this Pyrenean Provence phase, the Mesozoic sedimentary cover peeled off above the Triassic formation which played a role of soap layer. An important phase of folding prestructured the whole Provence and gave birth to the future massifs of Sainte-Baume, Sainte-Victoire, Etoile, Nerthe. The shortening observed during this Pyrenean Provence compressive phase formed anticlinal ramp structures overthrusting to the north while; nowadays, the surrection of the Alps causes a migration of the northernmost units toward the south.

2.2.5 Present Structure

The accepted structure for Basse Provence includes three main structural units overlapping from south to north (Gouvernet et al. 1979) (Fig. 2.11).

- The small Bandol unit consists mainly of Jurassic and Triassic terrains and overthrusts the Beausset unit to the north.
- The Beausset unit includes the Oligocene basins of Marseille and the important Cretaceous syncline of Beausset. It overthrusts the Arc unit at the level of the South-Provençal front.
- The Arc unit, a large syncline, of Late Jurassic to Eocene formations is in abnormal contact with the north Provence zone.

The complex tectonic history of this sector broke up or superimposed the various limestone units, which makes it possible the circulation of groundwater between the different units.

Port Miou and Le Bestouan springs are located at the foot of the Urgonian cliff (Fig. 2.2) that belongs to the western part of the Beausset Unit. The Urgonian (Barremian-Lower Cretaceous) is a massive limestone series whose thickness can reach several hundred of meters. In addition with the Jurassic limestone formation, this carbonate group is the most important karst reservoir in Marseille area. It is covered by the Aptian and Cenomanian (Middle Cretaceous) marls.

2.3 Climate

North of Cassis, the climate of Provence is of Mediterranean type with four seasons. It is rainy in autumn and spring. Rainfall is reduced with only 400 mm/y on the southwestern part, but it augments gradually to the northeast and to the east where it quickly reaches 1000 mm/y, in the Sainte-Baume or Sainte-Victoire massifs (Fig. 2.12).

2.4 Socioeconomic Context

Cassis is a small city of Provence, at the gates of Marseille, the third largest urban area of France (1,800,000 inh.). The theoretical watershed of the springs extends under several communes of Basse Provence, between two other large cities, Aix-en-Provence and Toulon. It is a territory with a high population density. Nowadays, the water supply of this area is mainly done by the aqueducts and pipelines of Canal

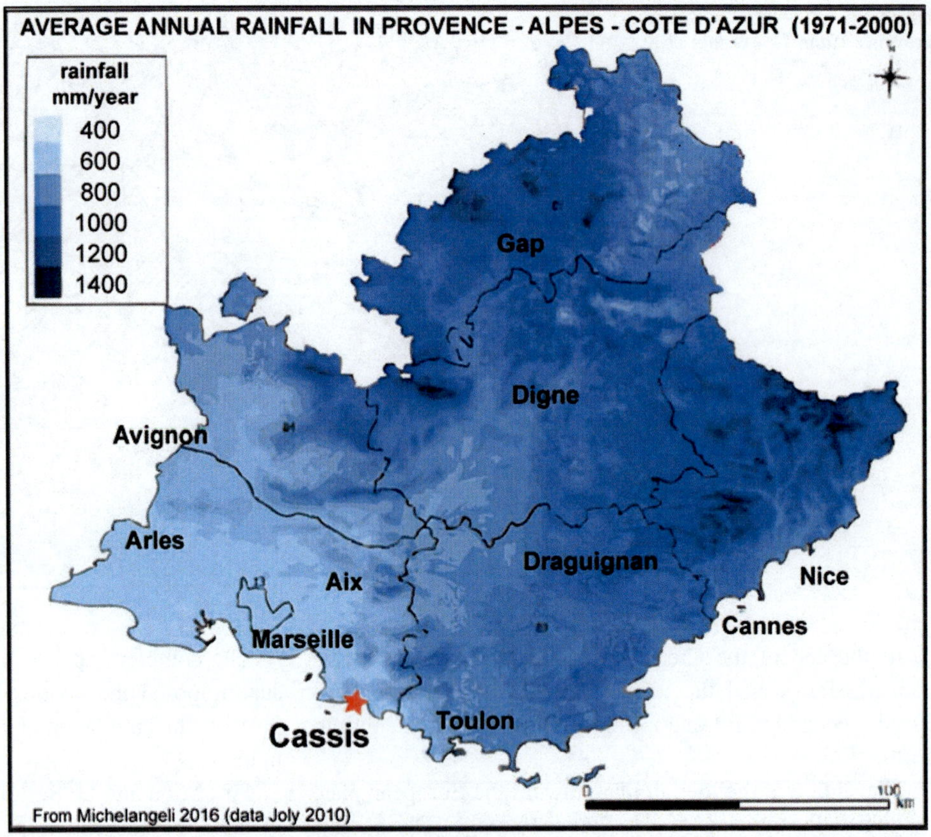

Fig. 2.12 Average annual rainfall in Provence—Alpes—Cote d'Azur (from Michelangeli 2016)

de Marseille and Canal de Provence societies. They catch the water from two rivers, the Durance (200 km long, 500 millions m^3/year) and the Verdon (600 millions m^3/year). Groundwater is poorly used.

The exploitation of Cassis groundwater is therefore an important issue since it could largely feed a large part of the coastal population between Marseille and Toulon. The attempts to catch Cassis underground river in the 1970s failed, but the aquifer that is drained by this karst system could be used by deep drillings and thus could constitute an alternative to the use of surface water. Some public or private wells are already soliciting it, such as the Coulin well in Gemenos or the Puyricard well in Cuges-Les-Pins, but no important work was realized.

References

Bestani, L., Espurt, N., Lamarche, J., Bellier, O., Hollender, F.: Reconstruction of the Provence chain evolution, southeastern France. Tectonics, American Geophysical Union (AGU) **35**(6), 1506–1525 (2016). https://doi.org/10.1002/2016TC004115

Gouvernet, C., Guieu, G., Rousset, C.: Provence, p. 238. Masson edit, Paris (1979)

Michelangeli, J.: Climatologie générale. Climat 3D. https://climat3d.hypotheses.org/ (2016)

Tassy, A.: Karsts côtiers et canyons sous-marins de la marge provençale au Cénozoïque: Contrôle géodynamique, eustatique, hydrologique et structural. Thesis geol., University of Provence, Marseille, p. 416 (2012)

3 Evolution of Explorations and Studies

3.1 Generalities

The flooded caves like Port Miou and Le Bestouan cannot be explored without diving equipment adapted to the underground world. The distances and depths reached during explorations are therefore closely dependent on the technological advances. In France, the first explorations were carried out at the Fountain of Vaucluse with copper-hat and heavy shoes by Ottonelli in 1878, but the use of the air hose prevented any significant progress in the networks. It is the invention of the "Le Prieur" autonomous scuba which allowed the beginning of real explorations. It should be noted that J. Y. Cousteau explored in 1946 at the Fontaine de Vaucluse where it reached 46 m depth, a value to compare with the 308 m depth currently known. He also explored Port Miou.

The modern great discoveries in deep caves are related to the use of gas mixtures, helium in the 1980s and hydrogen by the precursor J. Hasenmayer, which allowed him to reach alone, the depth of 200 m at Fontaine de Vaucluse in 1984. Today, the use of closed-circuit rebreathers is becoming widespread and cave divers easily overpass the kilometer: 5900 m at Doux de Coly (Dordogne), 3000 m at Bestouan (Calanques de Cassis) and depths beyond 100 m, the world record being −286 m at Font Estramar by the French diver Xavier Meniscus on December 30, 2019 (Bonzom 2020). He is the one who also reached the deepest part of Port Miou (see Fig. 6.2).

3.2 First Descriptions

The first written mention of Port Miou spring dates from Antiquity by the famous Greek geographer Pytheas (380–310 BC) who was living in Massalia (Marseille). The brackish water was probably not used at that time. It is necessary to wait until 1620 for the economic interest of the spring to be reported. The bishop of Marseille, Arthur d'Espinay de Saint Luc, asked King Louis XIII for the donation of a spring located in Port Miou creek in order to operate a wheat mill.

The first scientific description is attributed to the scientist Louis Ferdinand, Count of Marsili (1658–1730) in the page 13 of his book "*Histoire physique de la mer*" (physical history of the sea) published in Amsterdam in 1725. He describes the exact location of the main spring, the secondary outlets and the two coastal shafts. He draws a cross section of this subterranean and submarine river (Fig. 3.1) and presents the first theories on the origin of the water, from Sainte-Baume mountain and Cuges-Les-Pins area (Marsilli 1725). He attributes to the Romans the digging of both vertical wells. His hypothesis was that both wells were used for the water supply of Port Miou harbor, which is doubtful regarding the salinity of the river and the aspect of both shafts.

Then the spring fades into oblivion for nearly 150 years.

At the beginning of the twentieth century, Edouard Alfred Martel, the father of French speleology, was charged by the Ministry of Agriculture to consider a possible catchment of the spring, but after visiting the place he denied its existence (Martel 1907).

Shortly after, in 1908, the geologist Eugène Fournier wrote about the spring but he located it in the open sea, 300 m off the Pointe Cacau Cape (Gallocher 1954). Finally, Martel returned a few years later on his first conclusion and admitted the existence of the spring in 1930.

The first accurate localization work of Port Miou and Le Bestouan main springs and their secondaries outlets was published by Gallocher (1954). The location was confirmed later by conductivity and temperature measurements in Port Miou Calanques by Roques (1956). However, until that time the absence of diving equipment did not make it possible to have more information on the cave.

E. Gilli, *Port Miou and Le Bestouan (Cassis, France)*, Cave and Karst Systems of the World,
https://doi.org/10.1007/978-3-030-50192-1_3

Fig. 3.1 First cross section of the Port Miou cave by Marsilli (1725)

3.3 First Speleological Explorations

In the 1950s, the invention of the scuba diving equipment by Cousteau and Gagnan was an opportunity to explore the submarine galleries. The first known scuba diving exploration took place in 1953 by Eole group's cave divers from Toulon, but they did not go further than the two coastal shafts. Between 1955 and 1956, the French explorers H. Tazieff, J. Y. Cousteau and their team from the OFRS (French Underwater Research Office), made the first explorations. The network was explored over a distance of 240 m in Port Miou cave and 40 m in Le Bestouan cave (Corroy et al. 1958).

3.4 The SRPM Project

3.4.1 History of the SRPM Research Work

As soon as 1957, in search of a complementary water resource to the Marseille aqueduct, Cassis municipality encouraged hydrogeological studies on the submarine springs. Thus, during the summer of 1958, the geology laboratory of Saint Charles Faculty unsuccessfully conducted two surveys to reach the cave network of Port Miou and try to catch the water. The first attempt delivered brackish water and the other one failed (Corroy 1959).

In the 1960s, the population of Marseille was about 700,000 inhabitants and the demographic forecasts predicted a strong growth (2 million inhabitants were planned for the year 2000). This data and the presumed importance of these underground water resources for the Marseille region lead two organizations, the BRGM (Geology and nines bureau) and the SEM (Marseille water company), to study a catchment project. They joined in 1964 to create a research structure for Port Miou: the SRPM (Port Miou research syndicate).

The first objective of the syndicate was to reach the underwater gallery of Port Miou upstream of the Calanque entrance to develop a catchment project. Several campaigns of location and drilling took place. In 1964, the BGRM used various geophysical methods for determining the geometry of Port Miou cave from the surface (Cornet et al. 1963; Munck and Stanudin 1964). After this localization work, the drilling of a well was done in 1967 in Port Pin but it failed (Dellery et al. 1967). In 1966, a second attempt to reach the gallery was conducted using an already existing well in Port Pin Valley: the Germans' Well (Durozoy and Paloc 1966). This did not give a better result.

During the cave exploration in 1967 and 1968, the divers of the GEPS (Study and underground diving group) discovered a large air bell in the ceiling in the Port Miou gallery, 530 m from the entrance (GEPS 1968). This bell was named the "airbell 500." In 1969, a geophysical campaign using a magnetic dipole made it possible to locate the bell from the surface, in the Port Pin valley (Rouaud and Rodin 1969). The following year, a 45-m-deep well was hollowed and a 50-m-long tunnel was excavated to reach the airbell. This made it possible new explorations, studies and works in the cave.

3.4.2 First Phase of Acquisition and Exploration: 1968–1972

As early as 1968, the SRPM setted up a hydrometric monitoring of both submarine sources. Currentographers,

Fig. 3.2 Halocline in the entrance gallery of Port Miou. Upper part is not air but brackish water (*Photo* Swierczynski 2016)

recording the intensity and direction of the flows in the galleries, were installed into Le Bestouan then into Port Miou. In addition, samples of water were collected, more or less regularly, at different points of the system to determine the dissolved salts (mainly chlorides and sulfates). Finally, the speleological explorations completed the measurement system with visual observations.

Nevertheless, one of the major results of these first observations was the demonstration of a penetration of a seawater which formed a layer above the floor of the galleries of Port Miou cave network. This almost immobile water layer contributed to contaminate the spring water (Potié and Ricour 1971). The halocline, the contact between both layer is perfectly visible on the following picture (Fig. 3.2).

This observation was the beginning of a reflection to consider a project of dam with three objectives,

- Creating an impervious screen to prevent the intrusion of seawater in the gallery of Port Miou;
- Being able to set up a better hydrometric station (flow, chemical parameters);
- Catching the water to supply Marseille area.

An analogical modeling of the flow mechanisms of salt water and fresh water, according to the discharge variations of the river, was established (Thirriot 1971). It was done for establishing the dam characteristics necessary for a good separation of fresh water and seawater. A reduced model was then carried out at BRGM to determine the effectiveness of such a work in the Port Miou gallery (Ricour 1981).

The SRPM assisted by Coyne and Bellier, an engineering company, expert in dam construction, chose the solution of a "chicane" dam whose structure includes:

- A normal dam built on the floor of the gallery and closing partially the gallery;
- An upside-down dam hanging from the ceiling of the gallery, a few meters downstream of the normal dam.

This structure therefore blocked downstream the seawater in the lower part of the gallery and forced the low density fresh water, to circulate under the ceiling, over the seawater (Figs. 3.3, 3.4). In reality, the configuration of the gallery made it possible to avoid the building of the upside-down dam by using the natural shape of the ceiling. However, this part contained small galleries that could be hydraulic bypasses which had to be closed. This work was the first construction phase of the Port Miou submarine dam (Potié and Ricour 1974).

3.4.3 Implementation of the 1st Underwater Dam and 2nd Phase of Hydrometric Data Acquisition 1972–1975

The construction work for the 1st phase of the Port Miou submarine dam began in September 1972 at the "air bell 500" and ended at the end of that year (Fig. 3.5). The work was a world first and the SRPM and Coyne and Bellier appealed to multidisciplinary specialists to answer the numerous technical difficulties:

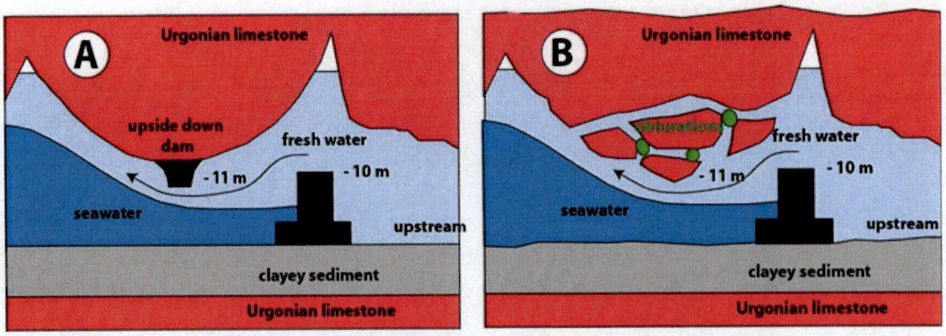

Fig. 3.3 Principle of "chicane" dam in Port Miou for blocking the seawater intrusion of seawater. **a** Model. **b** Local solution (Potié and Ricour 1974)

Fig. 3.4 Schematic view of the first dam in the Port Miou (Potié and Ricour 1974)

- Solétanche-Entreprise for the construction site;
- Comex and Hippocampe for the construction and the various underwater inspections.

The dam was directly founded on clay deposits lining the bottom of the gallery. In order to avoid any risk of subsidence, the grout used was a low-density one (d = 1.2 to 1.3). It was brought underground and underwater by a system of pipes. An ingenious and summary formwork made of metal tubes and onduline plates made it possible to quickly built the dam and thus limit the work of the divers (Potié 1974).

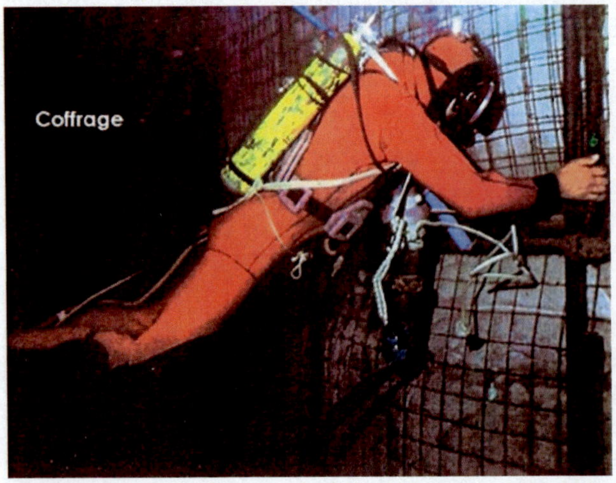

Fig. 3.5 Construction of the dam in 1973 (doc. SRPM)

In January 1973, the monitoring station located 150 m from the spring, hereinafter referred to as "the old station," was abandoned in favor of a more complete equipment in the "air bell 500" (Figs. 3.6, 3.7).

Blocking the intrusion of seawater was the main objective of the work. It was quickly reached. From the first weeks of recording, the seawater layer observed before the work in the lower part of the gallery, and responsible for a vertical salinity gradient, was no longer observed upstream of the dam. However, the chlorine concentration upstream of the dam remained higher than 3 g/L, even after several months of recording. This residual salinity indicated that there was a seawater contamination upstream of the dam at a great depth. The engineers supposed it came from a deep gallery at a depth around 80 m (Fig. 3.8). The construction of a higher dam was then planned (Potié 1974).

The first dam had been realized to block the marine intrusion, the principle of the second dam was to artificially increase the freshwater head. The objective was to lower the contact between fresh water and seawater for reducing the salinity (Fig. 3.9). This strategy is based on the Ghyben–Herzberg formula, which defines the position of the salt interface in a porous aquifer:

$$Z = \frac{\rho_f}{(\rho_s - *\rho_f)} \cdot H$$

where Z interface depth, H aquifer water head, ρ_f freshwater density, ρ_s seawater density.

Under normal conditions, Z is around 40 H.

The SRPM decided to use this principle in order to maintain this saline interface at a sufficient depth to improve the water quality of the spring. The project included a

Fig. 3.6 Monitoring station in the air bell 500 (doc. SRPM)

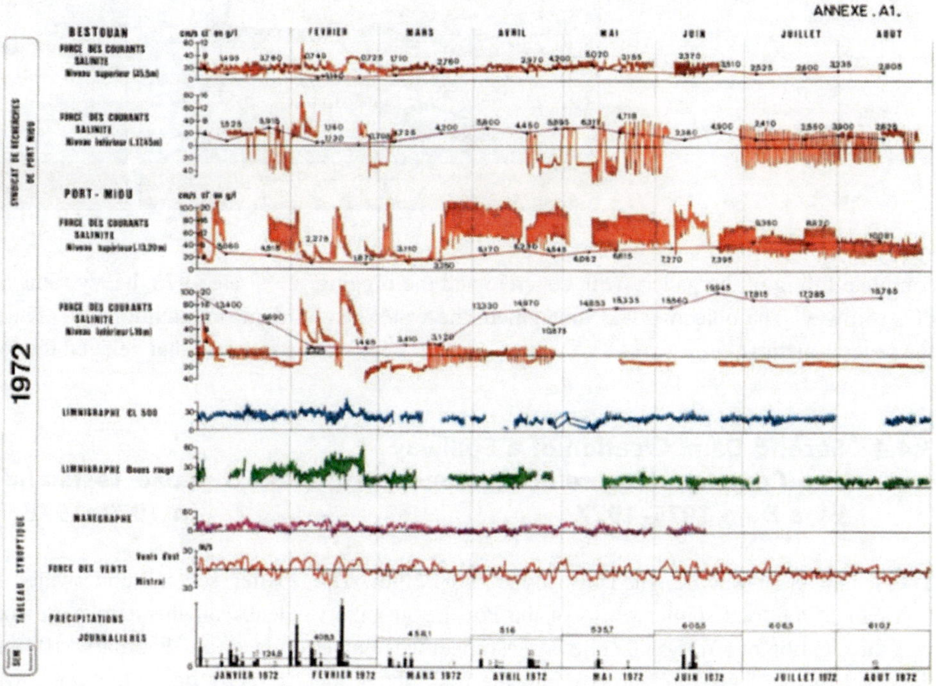

Fig. 3.7 Data sheet of the SRPM in 1972 (Potié 1974)

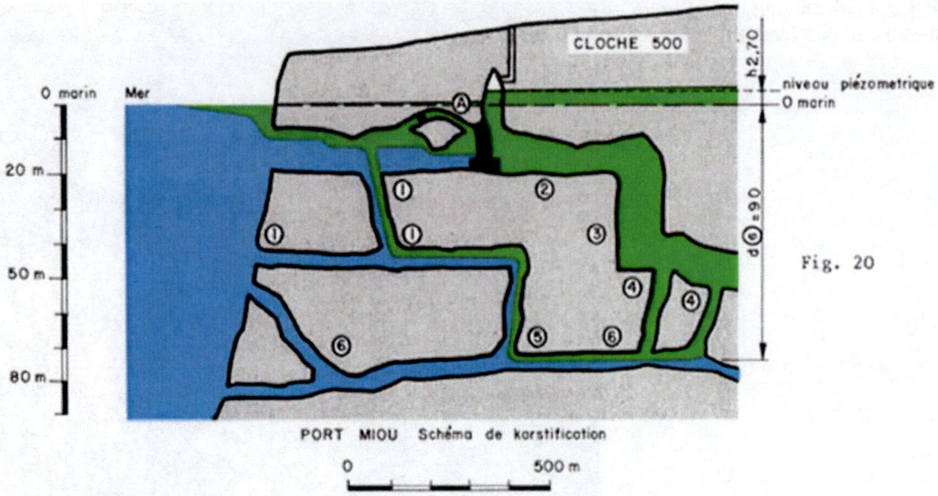

Fig. 3.8 Principle of a complete dam in Port Miou for blocking the seawater intrusion (Potié 1974). A 80-m-deep gallery was supposed to exist

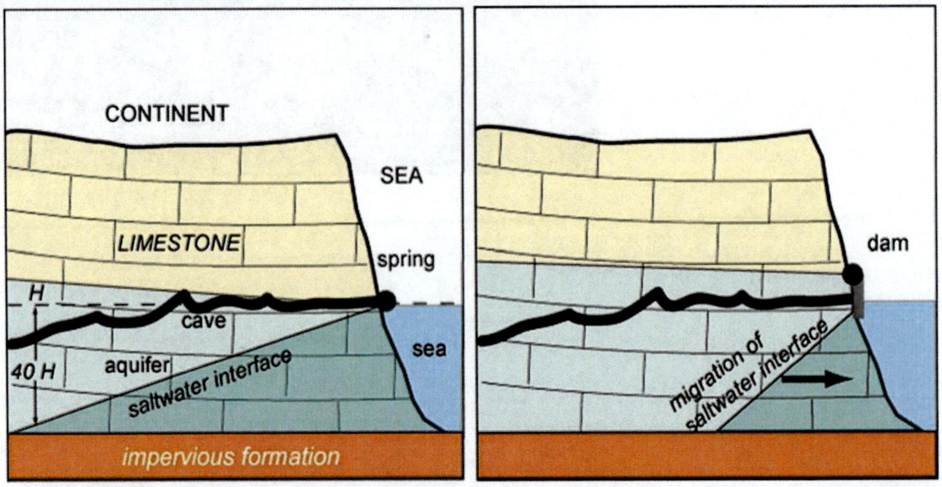

Fig. 3.9 Principle of water head augmentation to reduce the salinity (Gilli 2015)

complete filling of the gallery with concrete and the digging of a spillway. The objective was to augment the water head for several meters.

3.4.4 Second Dam: Creation of a Spillway and Complete Closure of the Cave by a Dam 1975–1977

The second dam was designed by Coyne and Bellier. The first part of the total shutter works of the Port Miou gallery was the creation of a free-spill spillway. A tunnel was dug at the level of the "bell 500" directly into the rock above the first dam (Figs. 3.10, 3.11, 3.12) between July 1975 and February 1976. Then as early as May, 1976, the building of the second dam was undertaken. The wall was built on the old existing dam. Four pipes were placed in the wall at a depth of −8 m NGF to make it possible controlled loading tests.

Late 1976, heavy rains affected the region and loaded the dam unintentionally. Numerous leakage problems were observed that delayed the loading tests until the beginning of 1977.

3.4.5 Load Tests and 3rd Acquisition Phase in 1977–1978

After several underwater sealing operations, the different leaks on the dam were fixed. The first loading tests took place on February 10 and 11, 1977. Balloons inflated directly under the water were placed at the entrance of the pipes, but they did not resist to the pressure and they burst. The test was then reported to the end of February. On February 26, a new attempt to obturate the pipes with wooden planks system was successful and the dam spilled for the first time. Unfortunately, the declogging of the clay, under the structure, created several important withdrawals.

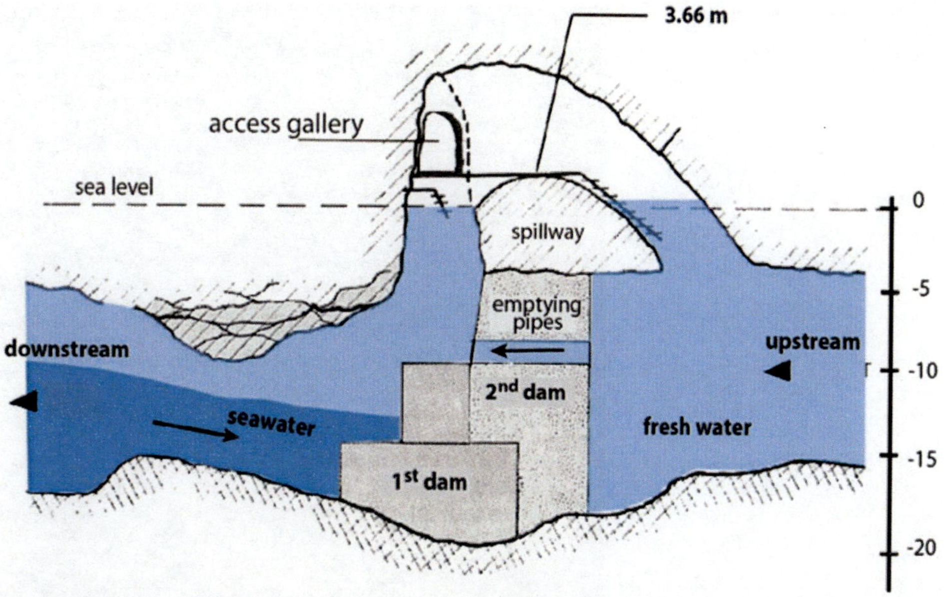

Fig. 3.10 Loading device project—schematic section of the Port Miou submarine dam (from Lombard 1977)

Fig. 3.11 Upstream wall of the second dam in 2005

Many underwater consolidation works were carried out until 1978.

In January 1978, a third loading attempt was made. Between 15 and 17 of January, rainfall on Marseille was nearly 200 mm and the spring experienced a sudden flood. On January 17, the water height above the dam reached 1.7 m which corresponds to a flow of 45 m^3/s (SRPM 1978). After this flood, the dam loading remained effective as the most important leaks had been repaired. The spill was flowing until June.

From 1977 to 1978, the SRPM continued to measure flow velocities, levels and salinity at both submarine springs, but the salinity remained too high, and finally, the site was abandoned in 1979 (Potié et al. 2005).

Fig. 3.12 Spillway in the air bell 500

3.5 End of the Hydrogeological Studies

After the failure of the project, the possible origin of the residual salinity, marine or continental, was studied. The SEM attempted a reconstitution of the water composition by mixing karst water with seawater, and it obtained the following result that shows a high sulfate content. The idea of marine contamination was retained (Table 3.1).

However, at same time, an isotope study on the sulfates (Vernet and Vernet 1980) subsequently showed a continental origin. It seems that this result has been extended to chlorides, thus providing a new explanation to Port Miou residual salinity.

That study probably contributed to stop the scientific studies on the site. However, the access to the dam facilitated the explorations. They allow the discovery by a Swiss team of an air bell and a deep shaft where M. Douchet dived down to a depth of 172 m in 1992 (Douchet 1993). In 2000, I started new studies on Port Miou based on these explorations and on new data I had collected on other submarine springs.

Table 3.1 Mains ions in Port Miou brackish water

Major ions	Concentrations
Ca^{2+}	100 mg/L
Mg^{2+}	25 mg/L
HCO_3^-	300 mg/L
Cl^-	30 mg/L
SO_4^{2-}	100 mg/L
Na^+	80 mg/L

References

Bonzom, N.: Pyrénées-orientales: en descendant à 286 m, un Français bat le record du monde de plongée souterraine. 20 minutes. https://www.20minutes.fr/planete/2690051-20200108-pyrenees-orientales-descendant-286-francais-bat-record-monde-plongee-souterraine (2020). Accessed 8 Jan 2020

Cornet, G., Durozoy, G., Gouvernet, C., Munck, F.: La source sous-marine de Port Miou (Calanques de Cassis). Etude par prospection géophysique. rapp. BRGM DS.63.A107. Orléans. p. 13, (1963)

Corroy, G.: Ville de Cassis - Etude d'une recherche hydrogéologique en vue de la réalisation d'une adduction d'eau complémentaire. Rapport interne, Laboratoire de géologie. Aix Marseille 1, 3 Nov 1959, p. 9, (1959)

Corroy, G., Gouvernet, C., Chouteau, J., Sivirine, A., Gilet, R., Picard, J.: Les résurgences sous-marines de la région de Cassis - La fontaine de Vaucluse. Résultats scientifiques des explorations de 1955 et 1956. Bulletin de l'Institut Océanographique **1131**, 1–35 (1958)

Dellery B., Durozoy G., Gouvernet C.: Sondage de Port Pin (commune de Marseille). Rapport interne, BRGM (DSGR 67 A47), 20 May 1967, p. 4, (1967)

Douchet, M.: Port Miou - Le Bestouan. Spelunca **49**, 5–6 (1993)

Durozoy, G., Paloc, H.: Projet de captage expérimental de la résurgence sous-marine de Port Miou. Rapport interne, BRGM (DS 66 A124), Dec 1966, p. 11, (1966)

Gallocher, P.: Contribution à l'étude sous-marine de Port Miou. Annales de Spéléologie - Spelunca **9**(3), 169–181 (1954)

Gilli, E.: Deep speleological salt contamination in Mediterranean karst aquifers: perspectives for water supply. Environmental Earth Sciences **74**(1), pp. 101–113, (2015)

Lombard, G.: Réalisation d'un barrage en galerie immergée à Port-Miou, pp. 3–1977. Cahiers des Comités de Prévention du Bâtiment et des Travaux Publics, Paris (1977)

Marsilli, L. F.: Histoire physique de la Mer. Amsterdam, La Compagnie edit. (1725)

Martel, E.-A.: Rapport sur un projet d'utilisation de la source sous-marine de Port-Miou, près Cassis (Bouches-du-Rhône), Ministère de l'Agriculture. Direction de l'hydraulique et des améliorations agricoles. - Paris: Impr. nationale Annales du ministère de l'Agriculture, **36b**, 11, (1907)

Munck, F., Stanudin, B.: Recherche des circulations karstiques par méthodes géophysiques électriques dans la région de Cassis (Bouches-du-Rhône). BRGM (DS64 A21), 22 Jan 1964, p. 8, (1964)

Potié, L.: Captage des résuregences sous-marines d'eau douce. Observations sur les effets du barrage expérimental de Port Miou, BRGM, Marseille, 74 SGN 272, p. 87, (1974)

Potié, L., Ricour, J.: Résurgences sous-marines de Port Miou - Cassis (Bouches-du-Rhône). *Rapport interne*, SRPM, 3 Dec 1971, p. 8, (1971)

Potié, L., Ricour, J.: Etudes et captage de résurgences d'eau douce sous-marines. Ressources en eau, pp. 5–18, (1974)

Potié, L., Tardieu, B., Ricour, J.: Port Miou and Bestouan freshwater submarine springs (Cassis-France). In: Stevanović, Z., Milanović, P. (eds.) Water Resources and Environmental Problems in Karst CVIJIC ' 2005, Spec. ed. FMG. University of Belgrade, pp. 266–274, (2005)

Ricour, J.: Construction d'un barrage d'essais souterrain à Port Miou (commune de Marseille) Bouches-du-Rhône. BRGM (Décision d'aide n° 76-7-1340 Mai 1981), p. 8, (1981)

Roques, H.: Localisation conductimétrique des émergences sous-marines de Port Miou. Annales de Spéléologie - Spelunca, t. XI (3ème série): pp. 109–112, (1956)

Rouaud, A., Rodin, G.: Localisation par la méthode du dipôle magnétique oscillant d'une cavité karstique à Port Pin - Port Miou (Cassis). Rapport interne, BRGM (69 GPH 037), 0ctobre 1969, p. 7, (1969)

Swierczynski, F.: Cave diving, Port Miou 2016 (U-W-X). https://vimeo.com/161296247 (2016)

Thirriot, C.: Etude schématique à l'aide d'un modèle analogique des écoulements en période de crue. Rapport interne, SRPM (C.T. N° 323), Oct 1971, p. 10, (1971)

Vernet, M., Vernet B.: Essai de discrimination par méthode isotopique de l'origine des eaux de systèmes karstiques. Application aux karsts continentaux et littoraux de Basse-Provence, Thesis Géologie Aix-Marseille 1, p. 208, (1980)

4 Resumption of Research in 2000

4.1 New Hydrogeological Studies

During the 1980s, I realized a large hydrogeological study in the area of Nice (South-Eastern France) where several submarine springs are present. A campaign of salinity and conductivity measures, coupled with GPS data, made it possible to locate and inventory these springs. The three main ones have an average flow around 500 L/s (Gilli 1999).

This study offered the possibility to equilibrate the local hydrologic balance. Indeed, the discharge of the inland springs was not sufficient regarding to the surface of karst land, but the presence of the submarine springs easily explained the observed deficit.

The water balance is easy to estimate as the discharge of most inland springs was studied since several years. The knowledge of both karst surface and annual spring discharge makes it possible to define a karst module **Mk** in L/km^2 for different geographic areas. For instance on coastal zone **Mk** = 5 L/km^2 while in higher elevation zone, Mk reaches 10–15 L/km^2. The following scheme presents the balance of the different karst units East of Nice (Fig. 4.1).

After the study of Nice springs, I was asked to make a similar work in Corsica and in Marseille-Toulon area where submarine or coastal springs were already known. Of course, among the different sites I focused on the Cassis springs. I had the possibility to collect an important part of the SRPM documents that had been preserved in the library of the DIREN PACA. Their analysis and my field observations in Nice area, trigged the start of new studies to explain the salt contamination in Port Miou (Gilli 2001) and the origin of the water.

4.2 A Marine Origin for the Salt Contamination

After the failure of the second experimentation of Port Miou, the thesis work of Vernet and Vernet (1980) had suggested that the salinity was not caused by seawater but by evaporites. I had important doubts on that explanation because other brackish submarine springs had been studied all around the Mediterranean Sea and for all examples, the salinity was only caused by a pure sea contamination. I decided to form a research team with Cl. Rousset (Marseille University) and B. Blavoux (Avignon University). The objective was to check if the salinity had a marine origin by collecting and analyzing water samples from different places:

- Pure seawater, collected far from the springs;
- Port Miou water, upstream of the dam, at different depths;
- Karst water in a deep well, a few kilometers inland, thus far from the sea.

The result was like expected. The salinity of Port Miou brackish water was only caused by a mixing of fresh water and seawater (Blavoux et al. 2004)

4.3 A Huge Aquifer to Explain the Discharge of Cassis Springs

In addition to my doubts about the origin of Port Miou stability I also had doubts about the size of the aquifer, estimated around 200 km^2. The study of the data acquired by the SRPM in the 1970s showed me that the discharge varied from 2 to 100 m^3/s, and the salinity ranged between 20 g/L during low water periods to 3 g/L during floods. I could estimate that the average discharge of fresh water was around 5 m^3/s for Port Miou.

A large watershed is necessary to explain this important discharge. A study of all of the karst units, north of Port Miou, showed that most of them have no springs or only low-discharge ones. This can be explained if the deep parts of these karst units contribute to the Port Miou aquifer. Analysis of the extent of limestone areas, the rainfall and the amount of water discharged by the springs, revealed a 50% global water deficit in the triangle Marseille-Aix-Brignoles.

E. Gilli, *Port Miou and Le Bestouan (Cassis, France)*, Cave and Karst Systems of the World,
https://doi.org/10.1007/978-3-030-50192-1_4

Fig. 4.1 Hydrologic balance in Nice area (Gilli 1999)

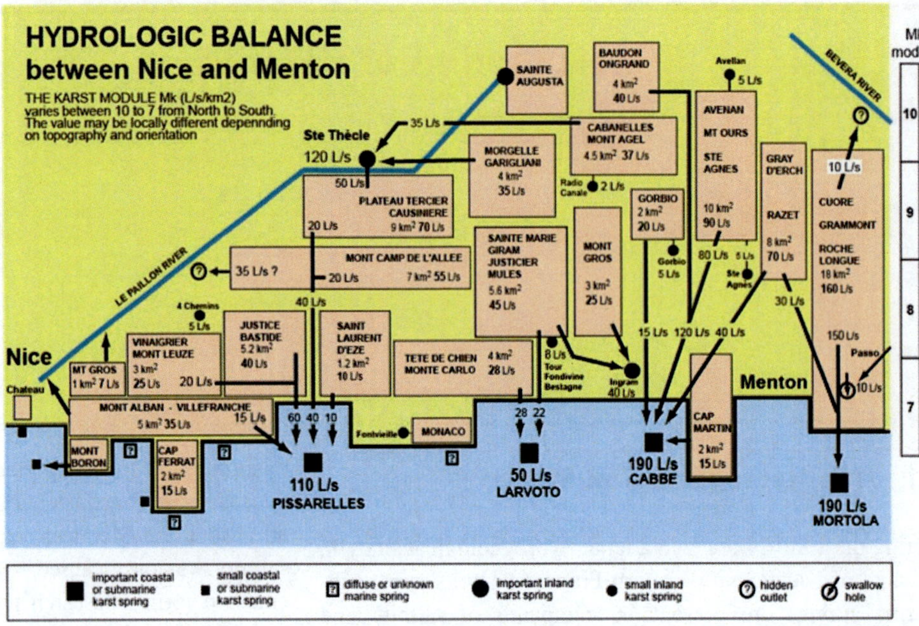

I proposed then a more than 2000 km^2 basin (Gilli 2001) to explain the size of the aquifer (Fig. 4.2).

These observations and the recent works on deep karst systems by Bini (1994) lead me to consider Port Miou as being an heritage of a big karst system that drained a large area of Basse Provence when the sea level was very low, probably during the Messinian (Gilli 2001). For such an assumption, I benefited of new data and concepts recently acquired and developed by cavers and karst scientists that are described below (Chap. 5).

Fig. 4.2 Supposed watershed (Gilli 2001, Blavoux et al. 2004)

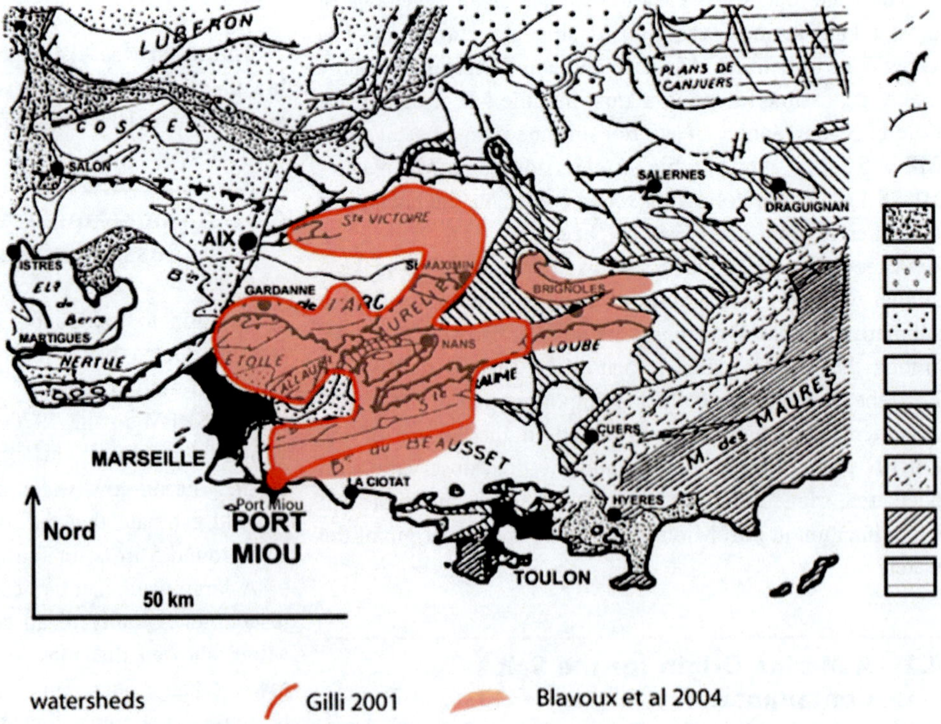

References

Bini, A.: Rapports entre la karstification périméditerranéenne et la crise de salinité du Messinien. L'exemple du karst Lombard. Karstologia **23**, 33–53 (1994)

Blavoux, B., Gilli, E., Rousset, C.: Alimentation et origine de la salinité de l'émergence karstique sous-marine de Port Miou (Marseille - Cassis – Bouches du Rhône). C. R. Geosciences, Elsevier, Paris **336**, 523–533 (2004)

Gilli, E.: Détection de sources sous-marines et précision de l'impluvium par mesure des variations de salinité. L'exemple de la source de Cabbé-Massolins (Roquebrune-Cap-Martin, France). CRAS Paris, IIa **329**, 109–116 (1999)

Gilli, E.: Compilation d'anciennes mesures de débit à Port Miou. Apport à l'hydrogéologie de la Provence. 7th coll. hydrogeol. en pays calcaire et milieu fissuré. Besançon, 20–22 Sept 2001, 157–160, (2001)

Vernet, M., Vernet, B.: Essai de discrimination par méthode isotopique de l'origine des eaux de systèmes karstiques. Application aux karsts continentaux et littoraux de Basse-Provence, Th. 3e cycle Géologie Aix-Marseille 1, p. 208, (1980)

5 Advances During the XXth Century

5.1 Exploration of Port Miou

During their explorations in Port Miou, the cave divers found a deep shaft at 2 km from the entrance. M. Douchet reached there a depth of 147 m in 1993, and with its powerful lamps, he was able to see that the shaft was much deeper, at least 180 m (Douchet 1993). The knowledge of this depth was important for two reasons:

- 180 m is higher than the lowering of the sea during the last glacial stages (−120 m) which suggested that the karst system could be a heritage from a period with a paleogeography very different from the current one.
- It is also more important than the vertical displacement of the haline interface provoked by the SRPM dam. Indeed, as the water head augmentation was 3.7 m, the Ghyben–Herzberg formula indicates that the fresh water/seawater interface is at a depth of 140 m which is less than the depth of the shaft (more than 180 m). This meant that the dam was not sufficient to prevent the deep sea intrusion (Gilli 1999) (Fig. 5.1).

5.2 Local Bathymetry

After the discovery of the prehistoric Cosquer Cave (Fig. 5.2), the geologist Colina-Girard (1992, 1996) published a bathymetric map that presents a karst plateau 150 m below the sea level. It is bordered on the east by a deep submarine valley: the Cassidaigne canyon. Both forms suggest a karst plateau cut by a pocket valley (Fig. 5.3).

Deep canyons are usually connected to a continental stream, but here, there is no river on the shore. Thus, my idea was that Port Miou river was the one that dug this canyon when the sea level was lower. During the previous glacial stages, the sea level was 120 m below the current one. This low level explains why prehistoric caves were not encountered before as our ancestors settlements are now located below the current sea level. The upper part of the Cosquer Cave, where the paintings were discovered, is an exception.

However, herein, the depth of the Cassidaigne canyon was too important to imagine a digging during the previous glacial stages. Indeed, the sea level dropped down to 120 m only. The recent description of a deep lowering of sea level during the Messinian salinity crisis gave therefore a perfect explanation to the presence of this deep canyon.

5.3 Messinian Evaporites

Messinian (Late Miocene) salt and gypsum layers are present on the border of the Mediterranean Sea. Different authors like Ruggieri (1967) proposed to explain their presence by a salinity crisis during which the sea evaporated and deposited these evaporite layers. They were also observed on the bottom of the sea during seismic surveying in the 1970s. Samples were also collected by the Glomar Challenger vessel during the Deep Sea Drilling Program in very deep zones. Ryan and Hsü (1973) assumed that they were the result of an important phase of desiccation of the sea and a level drop of 1500–2000 m.

5.4 Messinian Salinity Crisis

This part is an excerpt of a previous paper on Port Miou (Gilli 2015). During the Messinian, from 5.9 to 5.3 Ma, the salinity crisis (Hsü et al. 1973; Cita and Ryan 1978; Roveri et al. 2014) was a result of several successive drops in sea level, eventually decreasing to 2000 m below the present sea level (Fig. 5.4). The Mediterranean maintained exchanges

E. Gilli, *Port Miou and Le Bestouan (Cassis, France)*, Cave and Karst Systems of the World,
https://doi.org/10.1007/978-3-030-50192-1_5

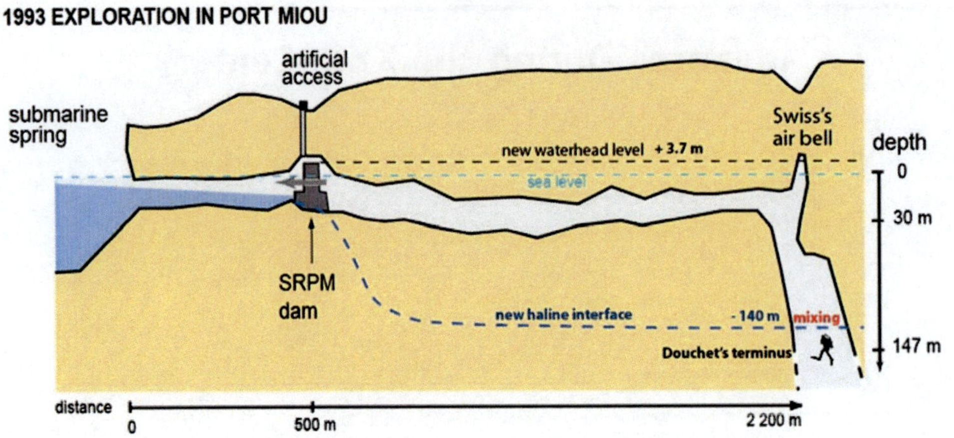

Fig. 5.1 New explorations in Port Miou in the 1990 (Gilli 1999)

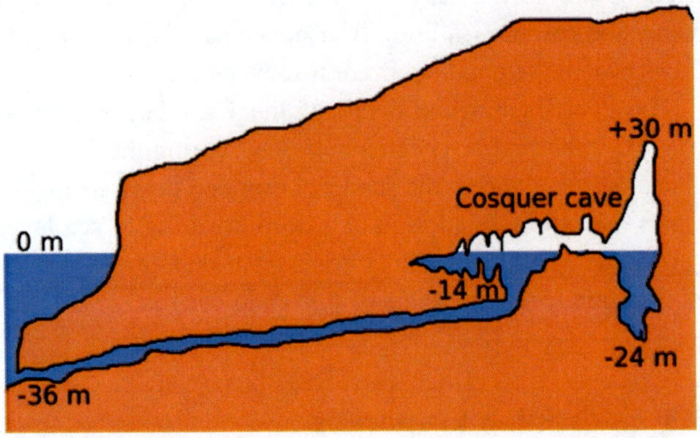

Fig. 5.2 Schematic cross section of Cosquer prehistoric cave

with the Atlantic through straights crossing through what is now southern Spain (the Baetian Mountains) and Morocco (the Rif Mountains) (Rouchy 1999). Inflow from the surrounding rivers was not enough to counteract the high evaporation rates in the Mediterranean basin. The series of drops in sea level modified the depositional environment, resulting in the deposition of thick layers of evaporites (gypsum and salt) in basin bottoms, as well as deep downcutting into the sedimentary layers along the continental shelf.

The major rivers, the Nile, the Pô and the Rhône, saw their beds cut downwards. The aquifers attached to these hydrologic base levels (sea, rivers or streams) also moved downwards. In the 1990s, this model of Messinian salinity crisis was widely accepted, and some authors had described their effects on coastal karsts (Bini 1994).

5.5 Consequences on the Mediterranean Rivers and Aquifers

The Mediterranean basin filled again, at the beginning of the Pliocene (5.3 Ma), due to the opening of the straight of Gibraltar (Roveri et al. 2014). The subsequent deposition of sediments was strongly influenced by the Messinian paleogeography. In Southern France, the sea filled the Messinian valleys, up to 80–100 m of elevation (Audra et al. 2004). This formed long rias that were gradually filled with sediment that blocked the karst water circulation. During the subsequent marine regression, these deposits were eroded which provoked new changes in the drainage of karst networks (Mocochain et al. 2016).

5.6 Application to Karst Aquifers and Cave Systems

Around the Rhône valley, these effects are perceptible for all the major karst springs, resulting in the formation of very deep conduits, fossilized passageways, pits, chimneys and deep Vauclusian springs (Audra et al. 2004; Mocochain et al. 2016) including of course Port Miou spring (Fig. 5.5).

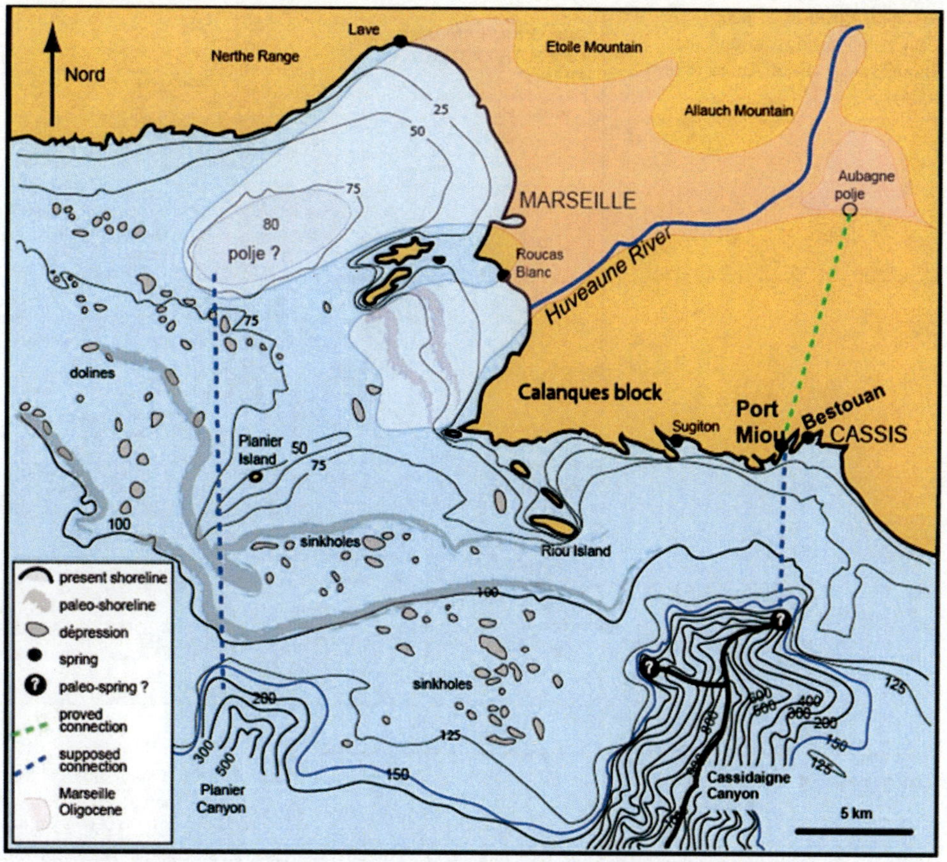

Fig. 5.3 Bathymetry map south of Marseille (from Collina-Girard 1996)

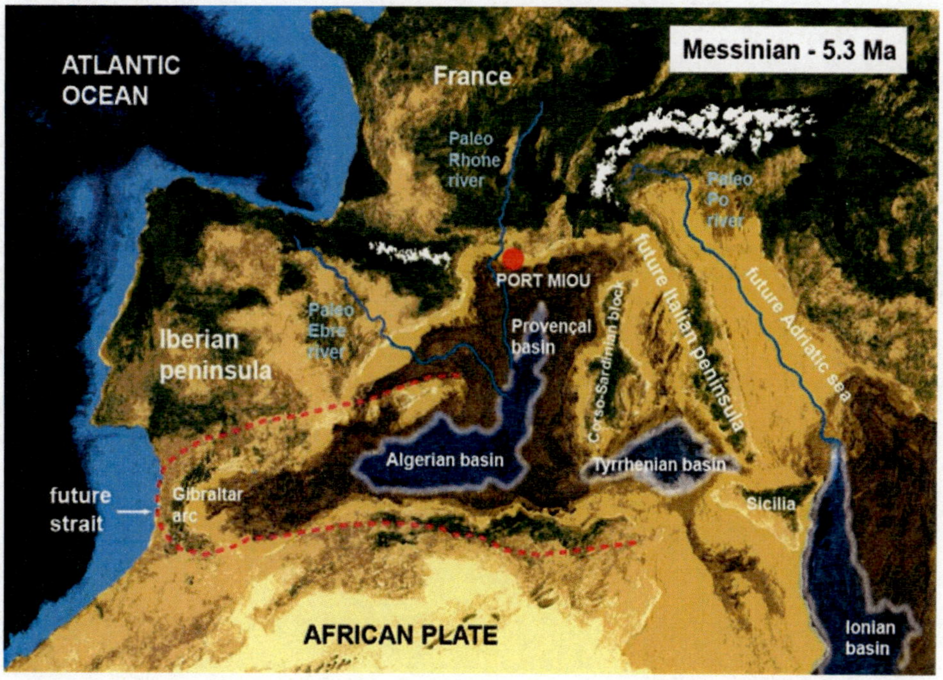

Fig. 5.4 Mediterranean Sea during the maximum of the Messinian salinity crisis

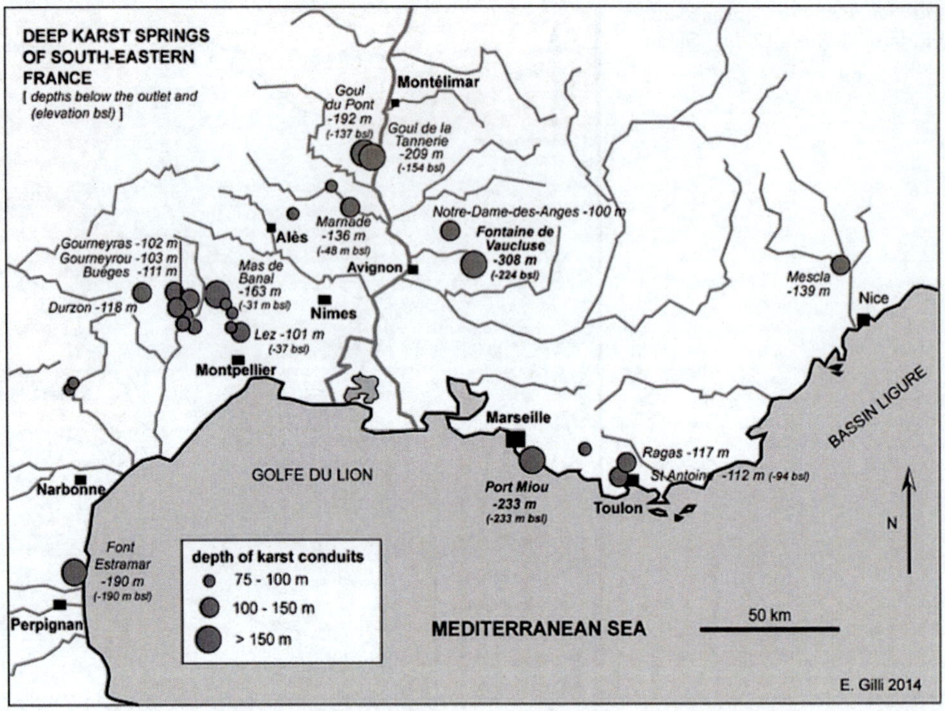

Fig. 5.5 Position of deep Vauclusian springs around the Rhône valley (from Audra et al. 2004)

References

Audra, P., Mocochain, L., Camus, H., Gilli, E., Clauzon, G.: The effect of the Messinian deep stage on karst development around the Mediterranean Sea. Examples from Southern France. Geodinamica Acta **17**(6), 27–38 (2004)

Bini, A.: Rapports entre la karstification périméditerranéenne et la crise de salinité du Messinien. L'exemple du karst Lombard. Karstologia **23**, 33–53 (1994)

Cita, M.B., Ryan, W.B.F.: Messinian erosional surfaces in the Mediterranean. Mar. Geol. **27**(3–4), 193–366 (1978)

Collina-Girard J.: Présentation d'une carte bathymétrique au 1/25000° du précontinent marseillais. (Au large de la zone limitée par la grotte Cosquer et l'habitat préhistorique de Carry le Rouet). Géologie Méditerranéenne. **XIX-2**, 77–87, (1992)

Collina-Girard, J.: Préhistoire et karst littoral: la grotte Cosquer et les Calanques Marseillaises (Bouches du Rhône, France). Karstologia **27**, 27–40 (1996)

Douchet, M.: Port Miou - Le Bestouan. Spelunca **49**, 5–6 (1993)

Gilli E.: Eaux et rivières souterraines, coll. Que Sais-je? n°455, PUF Paris, p. 128, (1999)

Gilli, E.: Deep speleological salt contamination in Mediterranean karst aquifers. perspectives for water supply. Environ. Earth Sci. **74–1**, 101–113 (2015). https://doi.org/10.1007/s12665-015-4042-2

Hsü, K. J., Cita M. B., Ryan W. B. F.: The origin of the Mediterranean evaporites. In: Ryan, W. B. F., Hsü, K. J. et al. (eds.) Initial reports of the deep sea drilling project 13 (1–2), US Government Printing Office, Washington DC, 1203–1231, (1973)

Mocochain, L., Clauzon, G., Bigot, J.-Y.: Réponses de l'endokarst ardéchois aux variations eustatiques générées par la crise de salinité messinienne. Bull. Soc. Geol. France **177–1**, 27–36 (2016)

Rouchy, J. M.: Un évènement exceptionnel: la crise de salinité messinienne de Méditerranée. In: Fröhlich, F., Schubnel, H.-J. (eds.), Les âges de la terre, MNHN, Paris, 104–108, (1999)

Roveri, M., Flecke, R. R., Krijgsman, W., Lofi, J., Lugli, S., Manzi, V., Sierro, F. J., Bertini, A., Camerlenghi, A., de Lange, G., Govers, R., Hilgen, F. J., Hübscher, C., Meijer, P. T., Stoica, M.: The Messinian salinity crisis: past and future of a great challenge for marine sciences, Marine Geology, **352**, 25–58, (2014)

Ruggieri, G.: The Miocene and later evolution of the Mediterranean Sea. Aspects of Tethyan Biogeography. Systematics Assoc. (London) **7**, 283–290 (1967)

Ryan, W. B. F., Hsü, K. J.: Initial Reports of the Deep Sea Drilling Project. (13), p. 1447, (1973)

6 A Messinian Model for Port Miou

6.1 Gilli's Model

These previous data and new concepts lead me to imagine that during the Messinian, the deep drop of the sea level provoked the lowering of the different karst aquifers of Basse Provence. The important depth of the lowering made it possible the hydraulic connection between different aquifers installed in different structural units. The drainage could then converge to a unique drain, the Port Miou river which was supposed to flow around 300 m below the present sea level. This river hollowed the Cassidaigne canyon and formed a karst pocket valley. Then after the Messinian crisis, the canyon was totally submerged, and Port Miou river found or dug another gallery to feed the current spring in Port Miou creek. The presence of this remaining Messinian low gallery could then be an easy way for a deep penetration of seawater into the karst aquifer (Gilli 2001; Blavoux et al. 2004).

This model had important strategic consequences. As we had proved that the salinity had a sea origin and that the aquifer was huge, the economical interest of Port Miou aquifer was again spotlighted.

I supposed that the seawater was arriving by a Messinian gallery located close to the end shaft of Port Miou cave. Thus, it looked realistic to find a place to catch fresh water by drilling in the basin far from that contamination zone. I convinced the regional water company (Société des Eaux de Marseille) to finance new research and I found an hydrogeologist student: Thomas Cavalera for a Ph.D. thesis.

6.2 Cavalera Thesis in 2004–2007

6.2.1 Objectives

The main objective of the thesis was to consolidate my model and to estimate the size of the basin by studying the effect of the rainfalls on the springs for an extended area. The method was as follow. On the one hand, collecting data on the spring (discharge, temperature and conductivity) and studying their variations. On the other hand, collecting information on heavy and short rainfall localized on different areas of the hypothetic watershed with the objective of finding a correlation between rainfall and spring variations. I had previously obtained very good results in Nice area, using that method (Gilli 1999). Both springs were monitored with velocity gauges and CTD probes, and data was collected from several weather stations (Figs. 6.1, 6.2, 6.3, 6.4 and 6.5). Unfortunately, the work suffered from three years of drought which did not make it possible to have a sufficient number of rainfalls. Thus, the PhD thesis ended without this part being developed, but valuable data was collected that are presented below (Cavalera 2007).

6.2.2 Hydrodynamics and Physico-Chemistry of Littoral Submarine Sources

Port Miou and Bestouan springs show a complex hydrodynamic behavior, linked to the combination of different physical and meteorological phenomena. The ocean tide is a condition at the hydraulic limits of the flow of the underwater sources by modifying the difference of hydraulic load between the sea and the underground gallery. This is reflected in the evolution of pressures and flow velocities that show significant semi-diurnal periodic oscillations correlated with the ebb and flow of the tide (Cavalera et al. 2006a).

The source of Port Miou has a very particular behavior demonstrated by the analysis of the phenomena occurring during its floods (Arfib et al. 2006). The mechanism of contamination of the karstic source with seawater is present throughout the chronicle. It does not seem to be affected by the successive loading of the system during floods because a minimum seawater flow is maintained, regardless of the floods studied. And the drops in salinity during floods are only related to freshwater inflows via ducts or fissures annexes before the terminal well.

E. Gilli, *Port Miou and Le Bestouan (Cassis, France)*, Cave and Karst Systems of the World,
https://doi.org/10.1007/978-3-030-50192-1_6

Fig. 6.1 Preparation of the monitoring in air bell 500

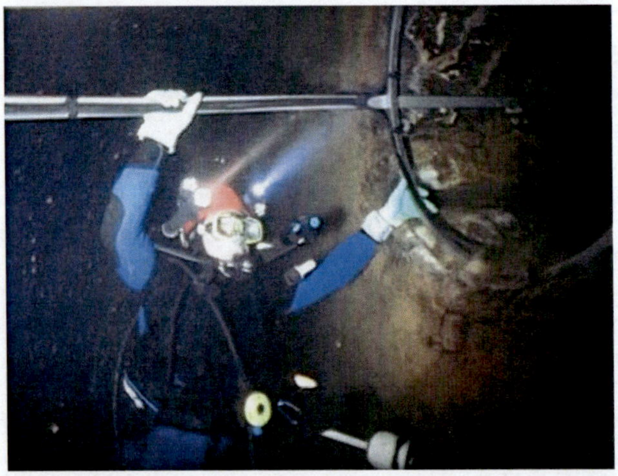

Fig. 6.2 Installation of a flow meter in the dam

Fig. 6.3 Flow meter in one of both large pipes that cross the dam

Fig. 6.4 Collecting data in the "air bell 500"

Fig. 6.5 Collecting data in the "air bell 500"

The general evolution of salinities and temperatures of both springs is correlated, with synchronous or near-synchronous falls, during floods, and increases, during low flows. There is, however, a physicochemical difference between the springs. The average salinity in Port Miou is almost twice as high as in Bestouan, with, respectively,

12.27 g/L for the former and 6.49 g/L for the latter, and the average water temperature in Port Miou is 0.8 °C higher than that of Bestouan.

The water analyses of the Port Miou cave confirm that the source is contaminated by seawater well upstream of the emergence over the entire tunnel to a depth of more than −170 m.

6.2.3 Research on the Watershed

Cavalera analyzed different karst units north of Cassis. On the one hand, he confirmed the observed deficit on the Jurassic–Cretaceous limestone massifs and on the other hand, the excess water status of the karst system of Port Miou (Fig. 6.6).

The particular structure of Basse Provence, despite a fragmentation into several well-individualized units, does not prohibit deep inter-unit hydraulic connections. The karst system of Port Miou is thus likely to be responsible for a deep drainage of neighboring limestone units that suffer a deficit. Thus, Cavalera proposed to extend the original 200 km^2 watershed to a more than 400 km^2 one (Fig. 6.7). This new basin implies that a large part of the infiltrations on the massifs of Sainte-Baume, Lare, Aurélien, Régagnas and south of Sainte-Victoire contribute to feed to the south the important submarine springs of Cassis.

The size of Cavalera's watershed is lower than the 1000 km^2 large one I previously proposed (Gilli 2001). This is probably caused by the rainfall deficit during his work. Discussions are still going on, but one thing is sure: The basin concerns an important part of Basse Provence.

6.2.4 Research on Geochemical Tracers of the Seawater Inflow

During Cavalera's thesis fieldwork, I made several dives upstream of the dam to install underwater monitoring equipment. This gave me a fantastic opportunity to support my model of sea contamination in which I assumed that seawater was supposed to reach the karst aquifer through a deep paleo-gallery connected to a spring located in the Cassidaigne pocket valley (Gilli 2001; Blavoux et al. 2004). We had called this model "the speleological contamination" (Cavalera et al. 2004; Gilli et al. 2004). The seawater discharge was later estimated 1 m^3/s (Cavalera 2007).

Indeed, while diving, I had observed a curious reddish layer on the floor of the flooded cave. Since the seawater inflow can carry tracers, strong consideration was given to this reddish level within the karst sediments of Port Miou.

This red level contains high concentrations of titanium. It is present up to a few centimeters below the surface of the sediment, but it is absent in deeper samples, which proves a recent contamination of the system. Titanium concentration is higher upstream of the dam, excluding contamination from the entrance of the cave (Fig. 6.8). Our hypothesis was that this strong geochemical signature of the titanium in the surface sediment could be the result of current anthropogenic inputs, which could be linked to the titanium-rich tailings that

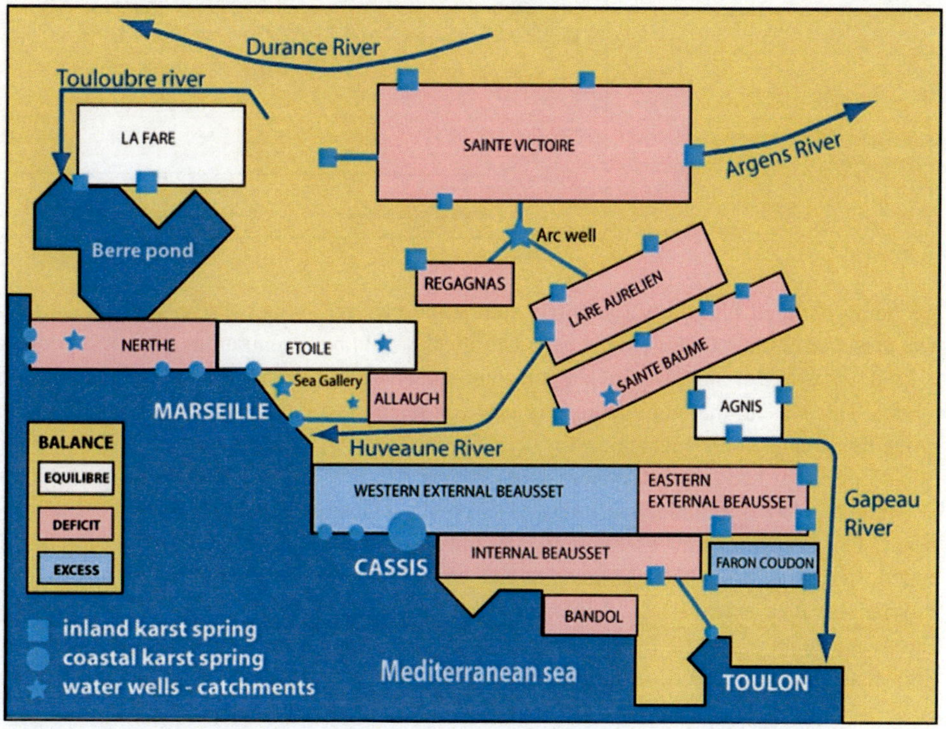

Fig. 6.6 Cavalera's hydrologic balance (from Cavalera 2007)

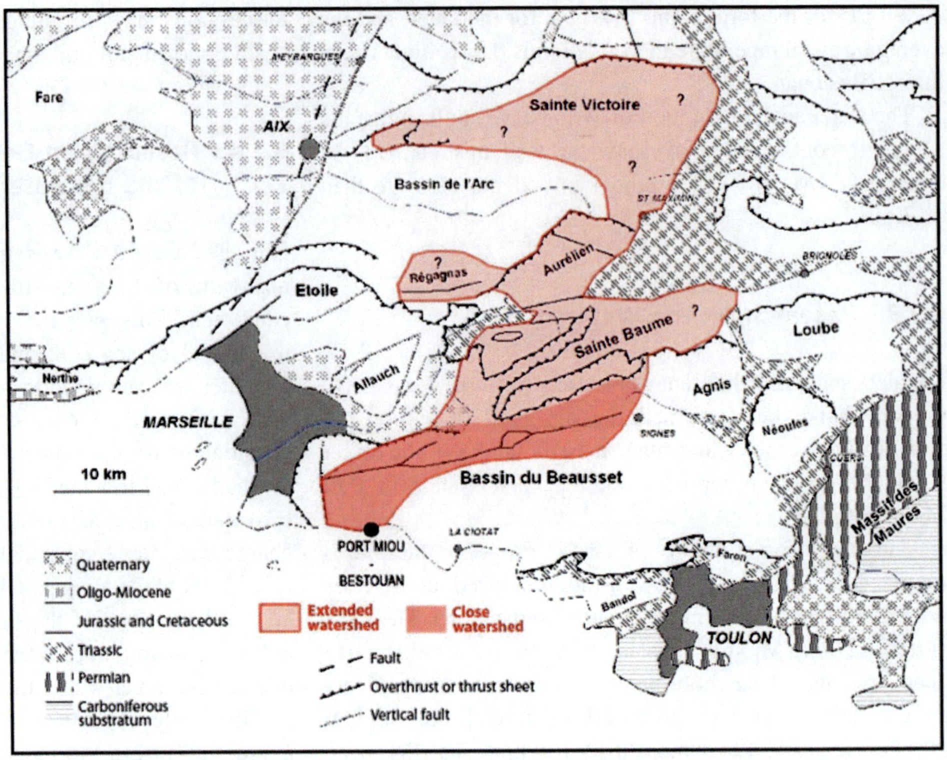

Fig. 6.7 Cavalera's extended or close basins (Cavalera 2007)

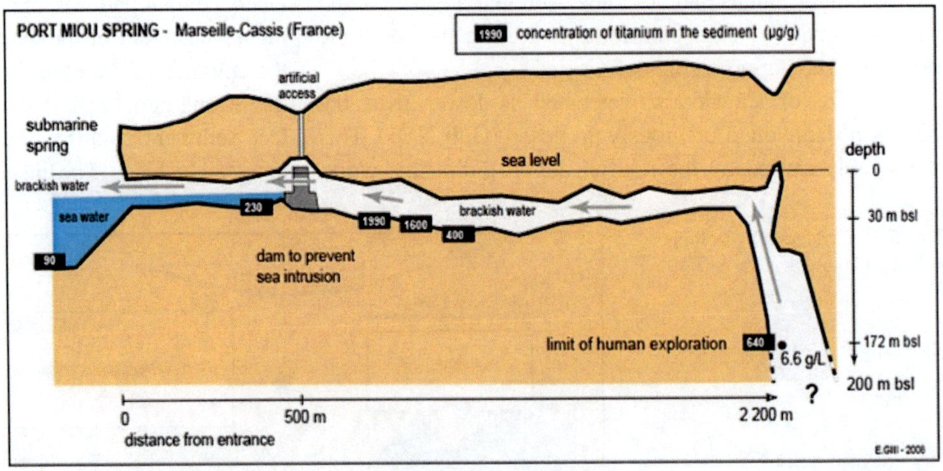

Fig. 6.8 Position of the samples in Port Miou and concentration in titanium. Note the important values upstream of the dam (Gilli 2015a, b)

are being dumped into the Cassidaigne submarine canyon. The presence of these tailings in the cave sediments would be a kind of dye test that would support our speleological model. This was all the more interesting since an attempt to prove this connection by searching for deep marine fauna in Port Miou water had failed (Cavalera et al. 2006b).

Since the 1970s, the alumina industry (Pechiney, Alcan, Alteo) has been dumping its titanium-rich tailings, locally called "red mud," into the Cassidaigne canyon. This deep submarine valley extends several kilometers off the coast of Cassis at a depth of 300 m (Blavoux et al. 2004; Bourcier and Zibrowius 1972). Red mud is a particularly fine mud (100% of the particles are smaller than 32 µm) that contains high concentrations of ferrous minerals (hematite, goethite and limonite) and various metallic oxides (TiO_2, MnO_2, V_2O_5…). It is deposited widely around the dumping site (Fig. 6.9). A scientific supervision committee (SSC) follows the impact of these discharges on the marine environment since 1995. Titanium concentrations, ranging between 500 and 1000 µg/g, have been measured in the Planier canyon's sediments, 20 km west of Cassidaigne canyon. This "red mud" plume, observed several kilometers from the dumping site, is a result of the horizontal transport of heavy metals (Arnoux and Stora 2003; Caors 1984) by the Ligurian Sea current.

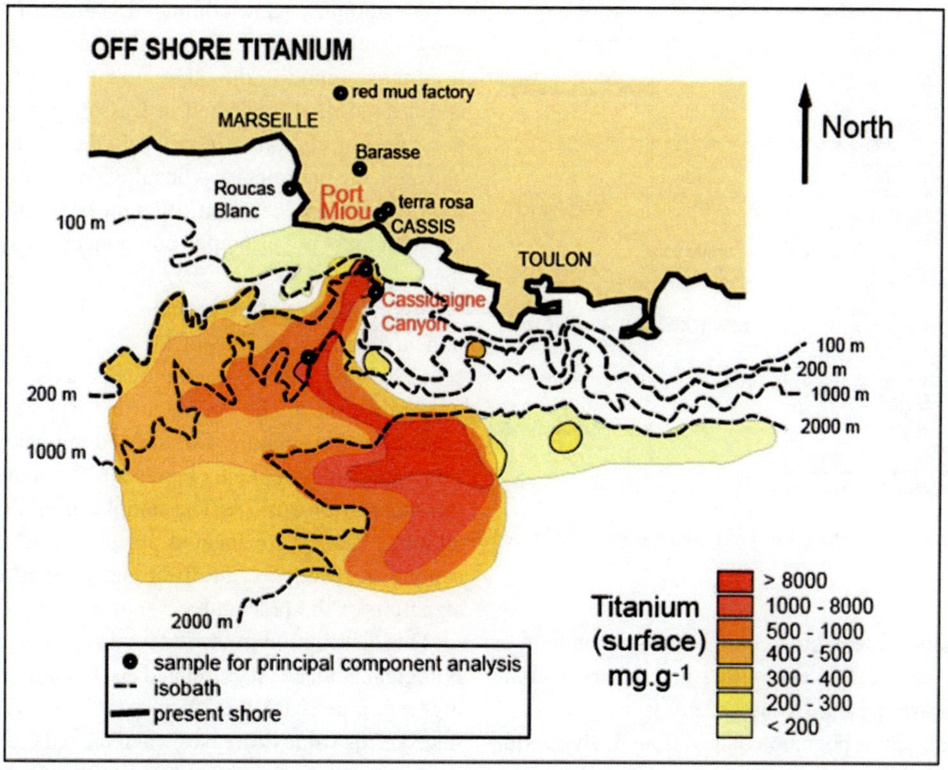

Fig. 6.9 Off shore titanium concentration (Gilli 2015a, b)

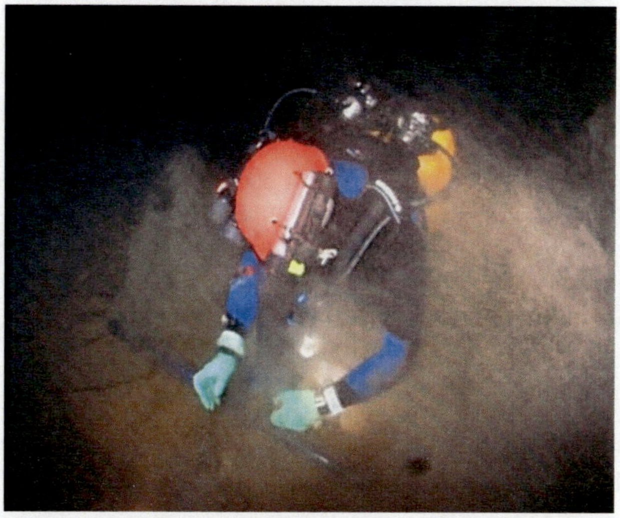

Fig. 6.10 Collecting mud sample in Port Miou

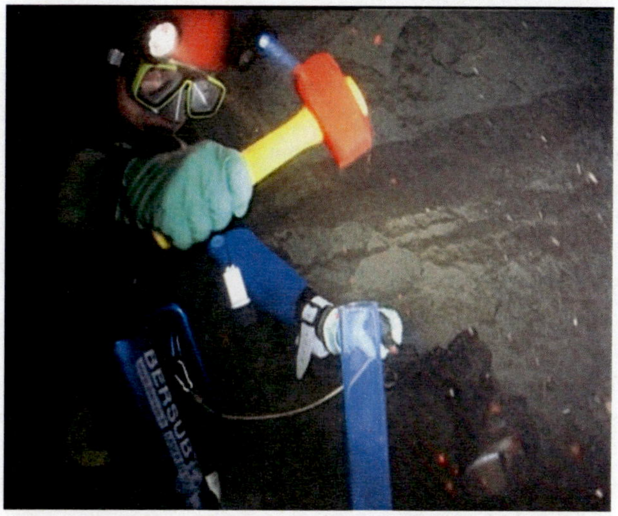

Fig. 6.11 Core sampling in the sediment of Port Miou upstream of the dam

Thus, the possibility that some trace metals could be sucked in and transported far inside a deep karstic conduit can be considered. The trace metal concentrations in the red mud constitute a chemical signature of an anthropogenic sedimentation. Samples of the karst sediment of Port Miou gallery were collected by cave divers, in three sampling locations at different places upstream of the dam (Figs. 6.10 and 6.11, 6.12).

6.2.5 Titanium Concentrations

Several local influences on coastal karstic springs must be taken into account to explain the presence of titanium:

- the contribution of the natural marine environment or Mediterranean influence;

Fig. 6.12 Core sample. Note the upper layer which is polluted by red mud

- the contribution of the natural continental environment or the influence of the Provence limestone where bauxite and terra rosa are present (Rousset 1968);
- one, or several, anthropogenic contributions from marine or continental origin.

To confirm the anthropogenic origin, the concentration of titanium in the sediment of Port Miou was compared with the already existing data and also with a sample of karst sediment collected in the Roucas Blanc, a coastal spring located in Marseille (Fig. 5.3). This spring is far from the red mud plume and from the dumping zone. The first analysis showed that the titanium concentration was 10 to 20 times higher in Port Miou than in the coastal sediments of Cassis Bay and almost 100 times higher than in the Roucas Blanc spring (Cavalera 2007).

To get more precise data, new analyses of heavy metals were conducted by LSRAE (Nice) (Cavalera et al. 2010). We collected samples in different places including Cassidaigne and obtained someones previously collected in Cassidaigne by the scientific supervision committee (SSC) of Alcan–Pechiney:

- red mud of Cassidaigne, that I collected in May 2009 at a depth of 300 m, close to the outlet, in its marine environment;
- sediments in Port Miou cave;
- sediment in Roucas Blanc spring;
- terra rossa (a kind of karst red clay) in a limestone crack above Port Miou river;
- sediment of Cassidaigne (Cassidaigne PU04 and Cassidaigne PU08) collected in 2007 by the SSC;
- red mud in La Barasse, a dumping site close to Gardanne;
- red mud from the Gardanne factory.

The samples were compared. Those of Port Miou cave show titanium and heavy metals concentrations that are far from those found in the other sites, except for the terra rossa and the red mud, collected in Cassidaigne in 2007, where a possible correlation is observed. As a simple comparison between the samples was insufficient, we decided to study the relationships between different heavy metals. A statistical approach by principal component analysis (PCA) was then conducted by the LSRAE (Cavalera et al. 2010).

Figure 6.13 synthesizes the results. The first two components explain 85.75% of the variance of the original data set. The first principal component axis represents the Ti contamination of the sediment and can be considered as a geochemical signature for the red mud axis. The second principal component axis represents the contamination of the sediment by Ni and Zn. The samples from Port Miou and the alumina factory are located at either end of axis 1, representing, respectively, the best and the poorest geochemical signature of the red mud.

This distribution of the red mud signature along the X-axis is logical. Indeed, it is related to Ti concentration, and the highest value is for the "factory" where the red mud is produced. This value decreases when the red mud is dumped into the sea and then transported far into the Cassidaigne canyon.

On the one hand, the samples in Port Miou are closer in composition to the samples collected in 2007 by the SSC in Cassidaigne than to the sample of terra rossa. On the other hand, they are far to those collected in the alumina factory or at the outlet of the submarine pipe. Thus, what is observed in Port Miou is close to what is measured for the red mud plume in Cassidaigne. The correlation is good with Cassidaigne PU04 and PU08. Both stations are located a few kilometers from the extremity of the pipe; thus, this observation is concordant if we take into account the fact that the surface sediment observed in Port Miou has also been transported for several kilometers in seawater.

This analysis reinforces the theory of an aspiration of seawater from Cassidaigne area to the deep parts of Port Miou cave. We named this model the "speleological model" (Fig. 6.14) (Cavalera et al. 2010), but it had to be supported by more precise studies to be totally proved.

6.3 Tassy's Thesis in 2009–2012

My previous works (Gilli 2001; Blavoux et al. 2004) and Cavalera's one (2007) suggested that the head of Cassidaigne canyon is a karst pocket valley which is the result of an active karst connexion between the Cassis springs and the canyon during the Messinian period. Froget's (1974), Collina-Girard's (1992, 1996) and Cavalera's (2007) studies highly suggested that the head of the canyon was a limestone zone. In addition, Bourcier and Zibrowius (1972) observed

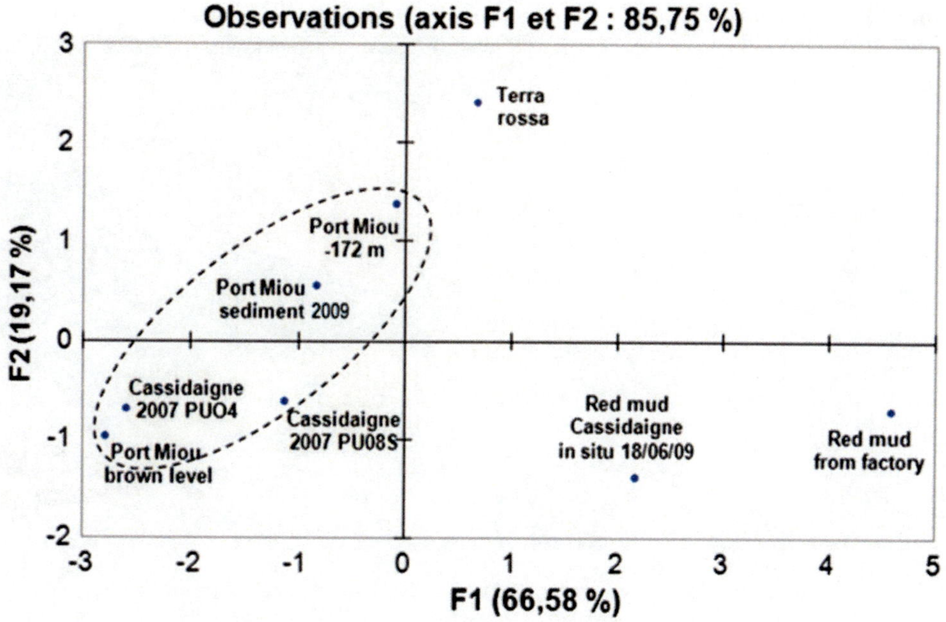

Fig. 6.13 First principal component analysis of the different samples that contain titanium and other heavy metals (Cavalera et al. 2010)

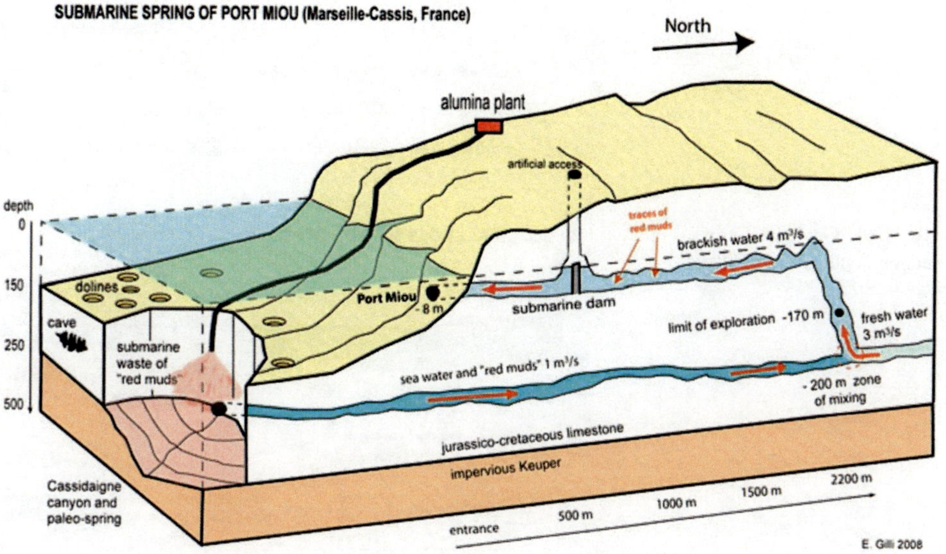

Fig. 6.14 Speleological model of contamination (Gilli 2011)

cave entrances and stalactites during dives with Cousteau's small submarines. However, this assumption suffered from the lack of geological data off Cassis and Marseille. To support the model, we had to definitely prove that the head of the canyon was a limestone area connected to the Calanques massif.

A geology thesis was therefore funded with several objectives:

- Confirming the continuity of limestone outcrops between the springs and the canyon;
- Studying the effects of the Messinian salinity crisis and the Pliocene transgression;
- Understanding the evolution and the functioning of coastal karsts and submarine canyons.

A Ph.D. student of Marseille University Aurélie Tassy (2012) conducted the research. Her work was based on the integration of geology, geomorphology and hydrogeology data. 2740 km of marine seismic profiles and 74 shallow corings were collected and studied. She made a new geological map (Fig. 6.15) and several land–sea geological cross sections (Fig. 6.16) of the continental shelf that extends from the western Nerthe massif to the Sicié Cape.

A new canyon, the «Bandol Canyon», filled with Pliocene sediments was discovered. It is oriented west–east and

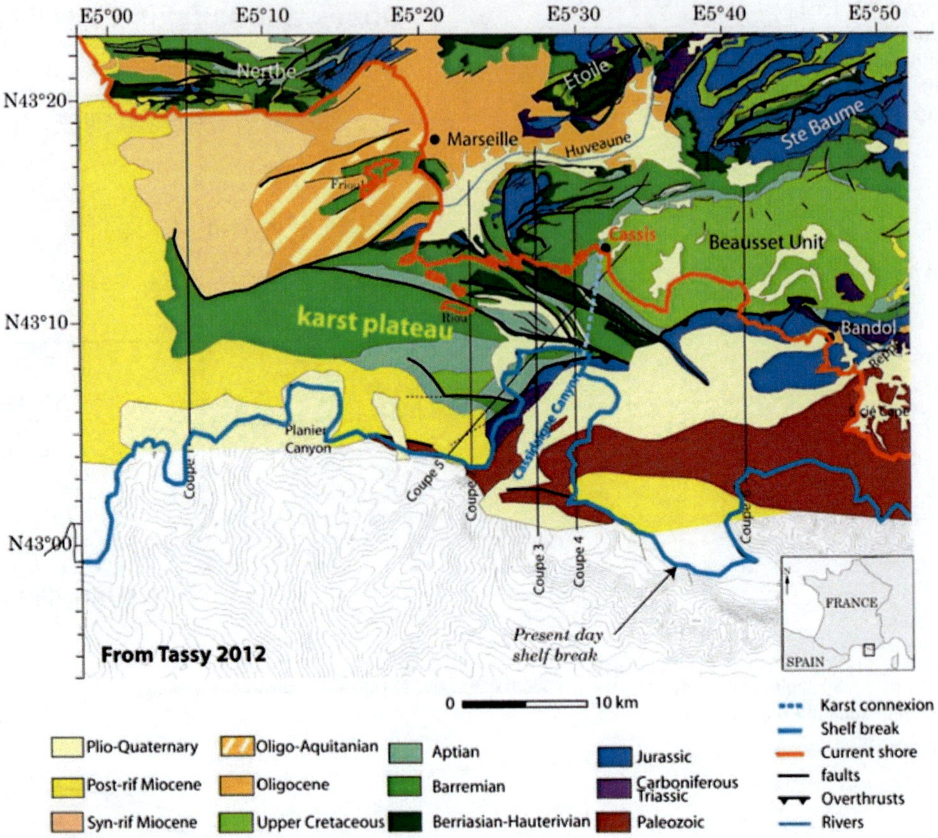

Fig. 6.15 Geological map off Cassis (from Tassy 2012)

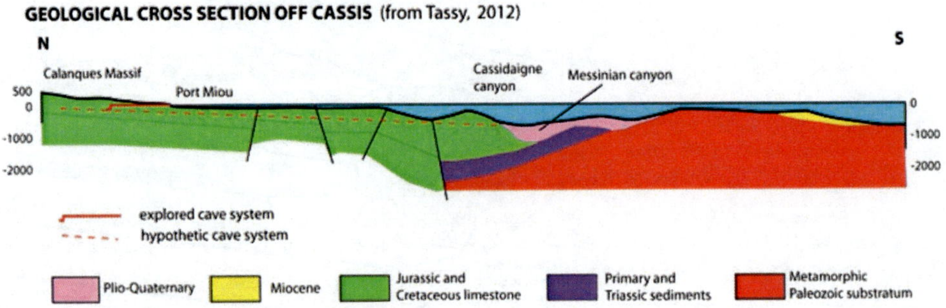

Fig. 6.16 Geological cross section south of Cassis (from Tassy 2012)

was mainly incised by the Reppe-Destel and Gapeau paleo rivers. The head of the Cassidaigne canyon is connected to it.

Limestone is present from Cassis to Cassidaigne which supports the hypothesis of a deep karst emerged during the Messinian with a pocket valley connected to Port Miou river cave. This karst plateau is confined by southern geological structures that are not favorable to karstification.

This work also gave valuable information on the paleotopography and the tectonic evolution of this area. Tassy also compiled hydrology data collected by Marseille University and the "Rivières Mystérieuses" group.

6.4 Different Models of Saline Contamination

6.4.1 Brackish Karst Springs

Most of the karst springs are brackish. The salt contamination occurs either at the entrance or in deeper parts. There are few examples of springs that deliver fresh water. It is usually only during flood periods, but the majority shows a salinity during low-water period that prevents their use for water supply. Several attempts were made to block the contamination, but they usually failed like for Port Miou (Gilli 2015a, Gilli et al. 2012). The Anavalos in Kiveri (Greece) is probably the only example where a circular dam protects the spring from the sea contamination. Understanding the mechanism is therefore a challenge. In Port Miou, several possible causes were reported or conceptualized.

6.4.2 Contamination by the Cave Entrance

The first explorers of Port Miou observed an almost motionless layer of seawater entering in the cave, close to the floor, below the brackish river flow. They used this sea layer to facilitate the dives toward upstream while they used the brackish current to go downstream. The first dam project was conducted to block this sea penetration that was supposed to be the main cause of salt contamination. The dam perfectly blocked the marine intrusion, but the water remained brackish. The failure of this first dam showed that the contamination was already acquired upstream.

6.4.3 Aspiration by the Bestouan

Some divers related a sea aspiration in the Bestouan cave, but no data confirms that affirmation.

6.4.4 Presence of a Gallery at −80 m

After the failure of the first dam, the idea of a deep gallery connected to the sea at a depth of 80 m was advanced. I do not know why that 80 m depth was proposed. I suppose it was defined in relation to what was known of the sea lowering during the last glacial maximum when the sea level was 120 m below the present one.

The aim of the second dam was to rise the freshwater head to lower the haline interface below that depth. The Ghyben–Herzberg formula predicted that the water head augmentation should be more than 2 meters to lower the haline interface at a depth higher than 80 m. The dam was 3.7 m high which would theoretically lower the interface down to 140 m. Thus, the discovery of the vertical shaft explored to a depth of 147 m in 1993 explained the failure of the second dam (see Sect. 4.2).

6.4.5 Deep Diffuse Contamination

The research on the Heraklion Almyros spring (Arfib 2001) suggested that the salinity of this karst spring was acquired in a diffuse way by the passage of the drain through a salty limestone matrix (Fig. 6.17a)

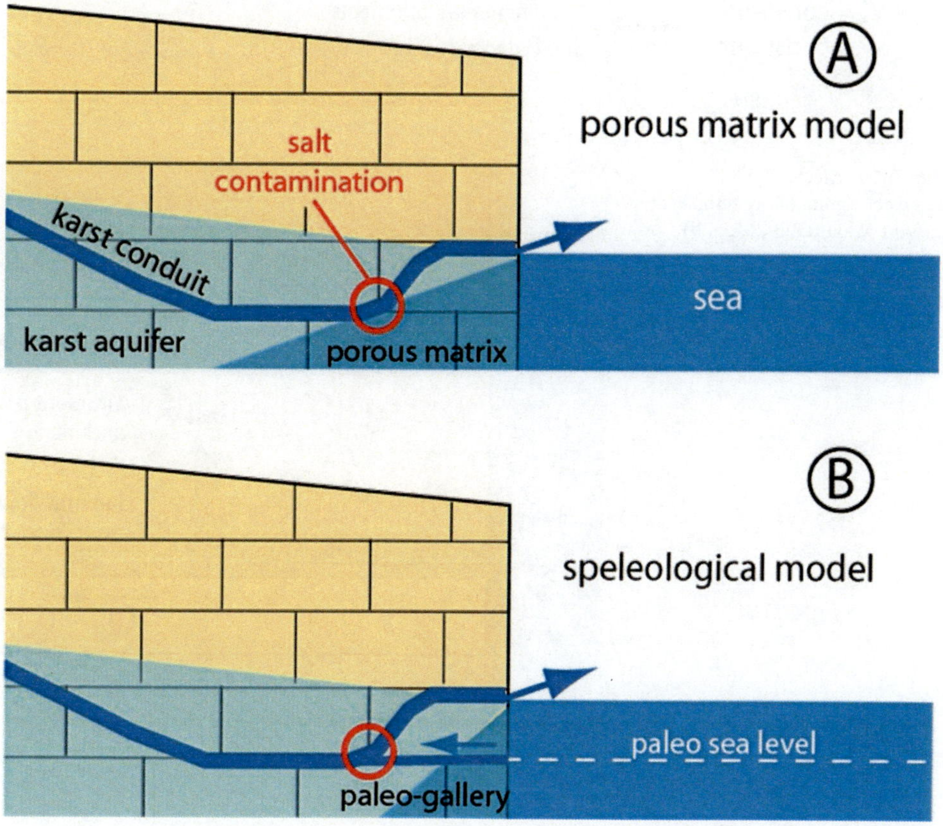

Fig. 6.17 Two models of salt contamination

6.4.6 Suction of Seawater by a Very Deep Paleodrain

As described above, the idea of a sea contamination by a side gallery connected to the main karst drain was advanced by several authors (Fig. 6.17b), but the mechanism of aspiration was not clear.

The classical mechanism involves a Venturi effect. However, such effect needs a specific geometry of the karst conduits, where a narrowing of the freshwater gallery is connected to conduit filled with seawater. The Bernoulli law easily explains the Venturi effect (Fig. 6.18). In accord with the principle of mass continuity, the speed of a water flow must increase as it passes through a constriction, while its static pressure must decrease in accord with the principle of conservation of mechanical energy. Thus, if a gallery is connected to a narrow place, seawater can be sucked.

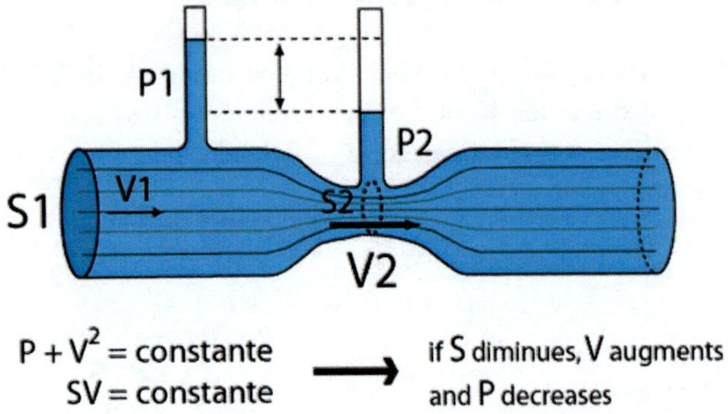

Fig. 6.18 Venturi effect

However, the volume of sucked seawater should be proportional to the freshwater flow augmentation which was not observed in Port Miou. Indeed, different tests (discharge versus salinity) show that the discharge of seawater is more or less constant around 1 m^3/s (Arfib et al. 2006; Cavalera 2007).

Thus, on the one hand, paleogeography, geology, presence of titanium, hydrometric data and cave diving suggested a deep gallery, but on the other hand, a Venturi effect was doubtful to explain a seawater suction.

I summarized the data and I proposed another process of mixing by turbulence, where the freshwater plume ascends in the seawater (Fig. 6.19). It is comparable to what occurs when a hot gas rises up into a chimney. It produces a draft of cold air that mixes with the very warm gas of the fire (Taylor 1945). In Port Miou, the turbulence at the contact between fresh water and seawater provoques a capture of seawater into the karst flow that becomes brackish. Seawater is withdrawn from the gallery, and, as this phenomenon takes place in a closed volume, it causes an aspiration of new seawater in the gallery. This mechanism explains the aspiration of seawater in the deep gallery, but it also explains the aspiration of seawater in the entrance gallery. But this only happens in turbulence regime; otherwise, the fresh water gently slips on the seawater. However, analyses of the current speed and the geometry of the cave show that the contact between both waters is turbulent most of the time (Gilli 2015b).

Fig. 6.19 Mechanism of seawater aspiration by turbulence in Port Miou (Gilli 2015a, b)

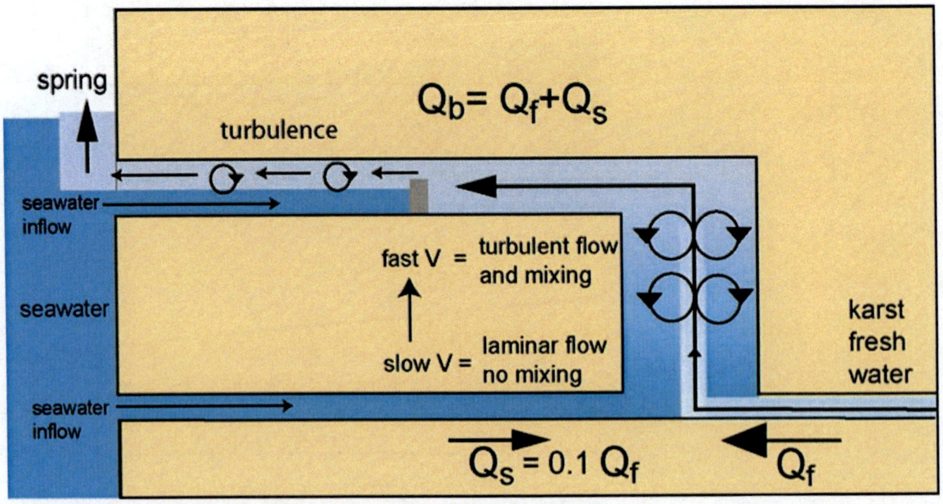

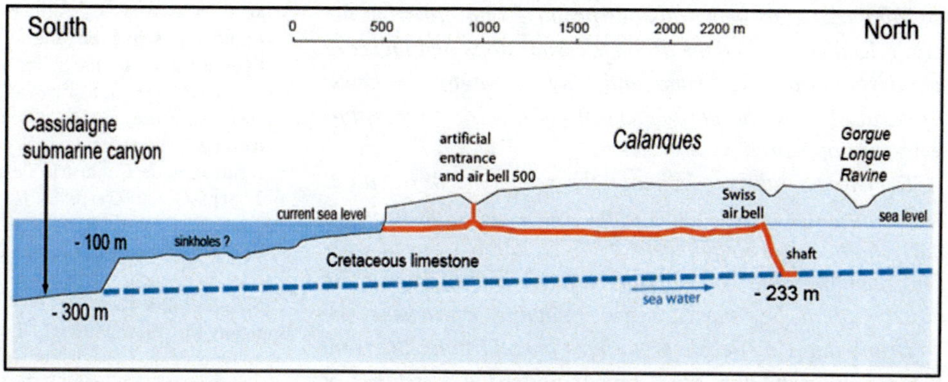

Fig. 6.20 Confirmation of the sea contamination by Meniscus in 2016

6.5 A Definitive Proof in 2016

On April 1, 2016, the French diver Xavier Meniscus reached a gallery junction at the depth of 232 m and crossed an halocline (Meniscus 2017) (Fig. 6.20). A gallery goes to the south, and the water is very salty which supports the hypothesis of a connection with Cassidaigne canyon. Indeed, Meniscus mentions:

> Diving exploration and deep gallery film by −232 m. After 20 m of exploration, arrival on a confluent. On the right (south direction), a big gallery that would go towards the sea. On the left direction north/east), a large diaclase that would go towards the basin. Strong halocline at background.

He was carrying a probe that indicates a salinity of 27–28 g/L on a 2–3 m thick layer of water in the terminal gallery, instead of 9 gm in the access gallery.

6.5.1 Xavier Meniscus Report

Friday 11 November

12 divers meet in the house at the entrance of the artificial pitch. Francis Schira installed a new faster winch for saving time when descending the diving equipment in the 44 m pitch that leads to the artificial gallery that gives access to the dam site. We carry down the equipment and prepare the exploration that will take place the next day.

Saturday morning

8:45 am: immersion from the dam, 530 m upstream of the Calanque entrance. I am equipped with two JOKI rebreather (mCCR) with a 12 L twinpak cylinder of 6/82 trimix, two 3 L oxygen cylinders, a 4 L air cylinder and two scooters.

Bruno, Guy and Manu will accompany me to the terminal shaft. Then, they will wait for me there until I come back from the bottom of the shaft to assist me during the deep decompression stop at −60 m.

A CTD probe is installed on my cylinders to record the salinity variations during the dive.

After 40 min of horizontal progression on a distance of 1700 m, we arrive at the head of the dialase shaft (point 2300 m).

I leave my main scooter at −18 m and take the BONEX Reference scooter which was specially prepared for a deep dive. I switch both rebreathers on the bottom gas and I start the descent into the first pitch, filmed by Bruno and Guy. I swim horizontally at −70 m, close to the ceiling, to reach the "Stanton" pendulum, then I descend slowly down to −130 m where I leave my 3L oxygen cylinder equipped with a Kiss valve limited in depth. I switch to my micro O_2 valve and I continue to descend to the bottom of the huge fault shaft. At −179 m, I find my 2009's hose reel. I tie a new line, because that of 2012 was torn off by the floods, then I continue the descent to the east.

At −220 m, I turn on the left, the diaclase ends with a huge giant's kettle. At the bottom is the entrance of the gallery explored in 2012. I enter in it. I cross a halocline, I overpass my 2012 terminus and explore a new horizontal gallery at the depth of −233 m. I advance in the gallery for 60 m. It is a flattener whose floor is covered with a small layer of sediment. The dimensions are 20 m wide by 4 m high. The cave continues horizontally to the north.

I turn back and start the ascent after a 6 min stay at the bottom. Deep decompression stops start around −165 m. Around −140 m, I see lights in the distance towards the ceiling. This is Manu who is measuring the dimensions of the diaclace with the depth sounder, in the −100 m deep zone.

At −65 m, at the top of the pendulum, I see the gallery discovered this year by Frédéric Swierczynski. A very high water current comes out with a slightly lower salinity. This income of less salty water raises new questions.

Around −60 m when I arrived in the last part of the shaft, I feel a dull pain close to my kidneys. It goes along my right leg, it comes with tingling and loss of sensitivity.

Notable. I start being anxious because I am far from the exit. I decide to extend the decompression stops and increase my oxygen window. Bruno and Guy arrive to get news. I inform them of the parameters of my dive and the symptoms that appeared.

Decision is taken to enhance the symptoms observation and to prepare an eventual round trip to the dam to bring back the necessary equipment if the situation were to worsen. Stéphane and Nico arrive to finish the survey of the diaclase shaft they started at the beginning of the year.

Bruno and Guy decide to go back for informing the team that stayed at the dam site. I remain under the assistance of Manu who ends at the same time its decompression stops.

Around −30 m the pain in my leg completely disappears which allows me to continue the decompression stops with a better mind. I will conclude the next day that it was a sciatica that will affect me for the following weeks.

My dive computers indicate that I can move up to −15 m. We still are at the point 2300 m, at −18 m deep and 1700 m from the dam where the decompression bell and the cylinders are. It's time to go home.

Manu accompanies me and we let Steph and Nico do their last measurements of the diaclase, in the so-called Swiss air bell. This work achieves the survey of the huge diaclase shaft down up to −100 m.

Forty minutes later, we are back at the dam. Having no more symptoms, I remain quietly with the rest of the team (...)

Finally, the last stop is displayed on the computer screen: 166 min at −6 m. I will do it in the decompression bell, breathing oxygen in open circuit from 3 15 L cylinder installed in the bell, with air rinses every 25 min. I will go into the bell with the assistance from Laurent who will help me get out and get in.

The bell is practical. It is warm, it is possible to eat, to drink and even to chating with the surface, using an intercom. I can eat nice food brought by my darling who stays with me.

During that time, the rest of the team evacuates the material which is no longer useful. There are at least two tons!

I reach the surface after a dive of 9 h 44 min with the help of Laurent and Eric who came to assist me and to recover the last decompression cylinders installed around the bell.

References

Arfib, B.: Ecoulements préférentiels en aquifères karstiques côtiers: impacts sur la salinité de l'eau dans le système de l'Almyros d'Heraklion, Crète, Grèce. In: Mudry, J., Zwahlen, F. (eds.) Actes du 7e Colloque d'Hydrologie en pays calcaire et en milieu fissuré, Besançon, France, 2001, pp. 13–16, (2001)

Arfib, B., Cavalera, T., Gilli, E.: Influence de l'hydrodynamique sur l'intrusion saline en aquifère karstique côtier (Influence of the hydrodynamic on the saline intrusion in coastal karstic aquifer) C. R. Geosciences 338, Elsevier, pp. 757–767, (2006)

Arnoux, A., Stora, G.: Analyses granulométriques et chimiques de 5 carottes de sédiments prélevés dans la zone de rejet des boues résiduaires de l'industrie de l'aluminium (Campagne ALPECAST 2, Sept–Oct 2002), p. 42. Rapp. int. Péchiney, Gardanne (2003)

Blavoux, B., Gilli, E., Rousset, C.: Alimentation et origine de la salinité de l'émergence karstique sous-marine de Port Miou (Marseille - Cassis – Bouches du Rhône). C. R. Geosciences Elsevier, Paris **336**, 523–533 (2004)

Bourcier, M., Zibrowius, H.: Les "Boues Rouges" déversées dans le canyon de la Cassidaigne (région de Marseille). Observations en soucoupe plongeante SP 350 (June 1971) et résultats de dragages. Téthys **4**(4), 811–842, (1972)

Caors, C.: Pollution minérale des sédiments entre le site de Cortiou et la Ciotat, p. 67. Thesis pharmacy univ, Marseille (1984)

Cavalera, T.: Etude du fonctionnement et du bassin d'alimentation de la source sous-marine de Port Miou (Cassis). Approche multicritère, p. 397. Thesis geol. University of Provence, Marseille (2007)

Cavalera, T., Gilli, E., Rousset C.: Hypothèse spéléologique de contamination saline de l'aquifère de Port Miou (Marseille). Actes 6éme symp. int. de l'eau, Cannes 2004, (2004)

Cavalera, T., Arfib, B., Gilli, E.: Ressource karstique côtière en Méditerranée: les sources sous-marines de Port Miou et du Bestouan (Marseille – France). 8e coll. hydrogeol. en pays calcaire. Neuchâtel, 21–25 September 2006, (2006a)

Cavalera, T., Chevaldonné, P., Gilli, E.: Speleological Hypothesis of Salt Contamination in the Aquifer of Port Miou (Marseille-Cassis, South of France) International Congress "Groundwater in Mediterranean Countries" (AQUAinMED), III International Symposium on Karst "Climate Change and Groundwater", annual meeting of the IGCP-513 Project of UNESCO. Málaga, 24–28, April 2006, (2006b)

Cavalera, T., Gilli, E., Mamindy-Pajany, Y., Marmier, N.: Mechanism of salt contamination of karstic springs related to the Messinian deep stage. The speleological model of Port Miou (France). EGU General Assembly 2009, Vienna, spec. issue. Geodinamica Acta **23** (1–3), 15–28, (2010)

Collina-Girard, J.: Présentation d'une carte bathymétrique au 1/25000° du précontinent marseillais. (Au large de la zone limitée par la grotte Cosquer et l'habitat préhistorique de Carry le Rouet). Géologie Méditerranéenne **XIX-2**, 77–87 (1992)

Collina-Girard, J.: Préhistoire et karst littoral: la grotte Cosquer et les Calanques Marseillaises (Bouches du Rhône, France). Karstologia **27**, 27–40 (1996)

Gilli, E.: Détection de sources sous-marines et précision de l'impluvium par mesure des variations de salinité. L'exemple de la source de Cabbé-Massolins (Roquebrune-Cap-Martin, France) CRAS Paris, **IIa-329**, 109–116, (1999)

Gilli, E.: Compilation d'anciennes mesures de débit à Port Miou. Apport à l'hydrogéologie de la Provence. 7e coll. hydrogeol. en pays calcaire et milieu fissuré. Besançon, 20–22 Sept 2001, pp. 157–160, (2001)

Gilli E., Mangan, C., Mudry, J.: Hydrogéologie. Objets, méthodes applications. Dunod edit. Paris. p. 303, (2004)

Gilli, E.: Karstologie. Karsts, grottes et sources. Coll Presup, Dunod édit. Paris, p. 256, (2011)

Gilli, E., Mangan, C., Mudry, J.: Hydrogeology, objectives, methods and applications, p. 394. Sciences publishers, CRC Press, Taylor & Francis Group, New York (2012)

Gilli, E.: Karstology. Karsts, caves and springs. CRC Press, Taylor & Francis Group, New York USA, p. 244, (2015a)

Gilli E.: Deep Speleological Salt Contamination in Mediterranean Karst Aquifers. Perspectives for Water Supply. Environmental Earth Sciences. (2015b) https://doi.org/10.1007/s12665-015-4042-2

Meniscus, X.: Encore plus profond à Port Miou: −233. Spelunca **146**, 27–29 (2017)

Rousset, C.: Contribution à l'étude des karsts du Sud-Est de la France: Altérations morphologiques et minérales, p. 523. Thesis geol University of Provence, Marseille (1968)

Tassy, A.: Karsts côtiers et canyons sous-marins de la marge provençale au Cénozoïque: Contrôle géodynamique, eustatique, hydrologique et structural, p. 416. Thesis geol. University of Provence, Marseille (2012)

Taylor: Dynamics of a mass of hot gas rising in the air. US Atomic Energy Comm. Report LADC-276, Los Alamos Scientific laboratory, p. 19, (1945)

7 Evolution of Explorations from the 1990s

7.1 Port Miou (from Acquaviva 2012)

7.1.1 Second Campaign of Explorations

In the 1970s, the construction of the underwater dam gave opportunities to cave divers to explore Port Miou upstream of the dam.

In 1978, P. Rousset and Cl. Touloumdjian explore the cave on 982 m; then, J. Cl. Dobrilla and B. Léger who were using underwater scooters reach 1165 m.

In June 1981, a team of Swiss divers (O. Isler, Cl. Magnin and P. Perracini) reach an air bell 2000 m from the entrance but do not find any continuation.

In July, the French caver B. Léger discovers a vertical shaft from which the water flow comes. He explores it until a depth of 82 m without seeing the bottom.

For different reasons, like administrative problems or technical difficulties, the explorations end for 11 years.

7.1.2 Third Campaign of Explorations

The use of gas mixture and the construction of small decompression bells make it possible to explore the end shaft.

With the assistance of the French diving company Comex, M. Douchet and his team reach the depth of 120 m in September 1992.

Two years later, he explores the shaft until its end at the depth of 147 m where he observes an inclined gallery.

Due to the long distance and the important depth, the explorations stop again for a decade.

7.1.3 Last and Current Campaign of Explorations

The use of rebreathers and gas mixtures offers new possibilities that were used in Port Miou from 2005.

In April 2005, X. Meniscus and J. Meynié and their team explore the shaft until 150 m.

In April 2008, the same explorers reach 178 m. Then, in 2012, X. Meniscus follows the exploration of the passage down to −225 m. The conduit which is inclined until 190 m is now merely horizontal.

The last explorations took place in 2016. Xavier Meniscus reached the depth of 233 m and explored a horizontal gallery for 60 m. The cave continues further.

7.1.4 The Use of Closed-Circuit Rebreathers

Breathing air brings into the lungs a gas mixture of 21% oxygen and 79% nitrogen. The latter, also called diluent, only acts as a transporter and is not be used by the body. Thus, on the one hand, nitrogen is not consumed and is rejected on inspiration in an amount identical to what is absorbed on inspiration. On the other hand, oxygen is consumed by the organism to fuel the metabolism. Each expiration rejects 16% of oxygen, while 21% is absorbed during each inhalation. In practice, this means that only 1/4 of the oxygen is breathed and 3/4 is lost when diving. In the blood, the oxygen that was consumed by the body's cells is replaced by a waste product: carbon dioxide (CO_2) that is extracted and then rejected by the lungs during the expirations.

A rebreather is a diving equipment that makes it possible to recycle the air during the breath. Only 5% oxygen is

E. Gilli, *Port Miou and Le Bestouan (Cassis, France)*, Cave and Karst Systems of the World,
https://doi.org/10.1007/978-3-030-50192-1_7

consumed, while the 16% oxygen is recovered. The carbon dioxide is captured in a filter thanks to lime which has the property of fixing the CO_2 while producing heat.

As nitrogen provokes diving narcosis, deep dives implicate to add helium as diluent.

In Port Miou, the use of closed-circuit rebreather (CCR) was essential, but the depth to overpass, −233 m, constitutes a limit which, if it is not absolute, remains difficult to exceed. More than 300 m depths are highly probable, and the use of a robot is unavoidable. The drilling of a well at the vertical of the Swiss air bell is planned. It would enable to install an umbilical wire for a remotely operated underwater vehicle (ROV).

7.2 Le Bestouan

At Le Bestouan, the long distances and the nature of the ends of the galleries, where narrow passages or unstable screes are present, no longer allow to consider classic diving explorations. Due to the difficulties of getting a precise survey, several magnetic beacons were installed in different places of the cave to make it possible a precise location from the surface and to estimate the possibility of finding other access either by a natural shaft (see 8.3) or by an artificial pitch. The idea of a large diameter drill to reach the most remote areas of the cave is advanced.

7.3 Synthesis of the Explorations

See Table 7.1.

7.4 Other Caves

Few caves, likely to be connected with the upstream of Port Miou and Le Bestouan, are known in the Calanques massif. However, one can mention five shafts (Fig. 7.1):

- Aven des Marseillais, which reaches the altitude of 47 m a.s.l. and whose deep parts are filled with water during floods (Guieu et al. 1996),
- Gouffre des Gorguettes. This shaft is near Mussuguet-3 shaft. It has been explored to the altitude of 71 m a.s.l. It shows traces of water loading (Guieu et al. 1996),

Table 7.1 History of Cassis cave exploration (from Acquaviva 2012)

Date	Divers	Port Miou		Le Bestouan	
		Length		Length	
1953	EOLE group	40 m			
1955	H. Tazieff	220 m	First dive		
1955	J. Y. Cousteau and OFRS	280 m		−15 m	
1965	GEPS Touloumdjian	350 m		120 m	
1966	C. Touloumdjian	400 m		400 m	
1968-1972	SRPM GEPS	870 m	Discovery of the air bell 500		
1976	FFESSM			470 m	
1977	COMEX			620 m	
1977	B. Léger			770 m	Cul de sac
1978	C. Touloumdjian			1300 m	
	J. C Dobrilla; B. Léger	1165 m			
1978	F. Le Guen			1400 m	
1981	Swiss cavers	2000 m	Discovery of an air bell		
1981	Bertrand Léger	2150 −82 m	Discovery of the terminal shaft		
1983	C. Touloumdjian			2050 m	
1989	F. Le Guen			2290 m	
1990	M. Douchet			2665 m	

(continued)

Table 7.1 (continued)

Date	Divers	Port Miou		Le Bestouan	
		Length		Length	
1991	M. Douchet			2950 m	Discovery of three impenetrable galleries
1992	M. Douchet	2190 m −123 m		2640 m	Discovery of Flou Gallery
1993	M. Douchet	2230 m −147 m			
1993	FFESSM Marseille			2950 m/-31 m (Grande Galerie) 850 m/-33 m (Galerie du Flou)	End on a scree on both galleries
2005	X. Meniscus	2230 m −150 m			
2005	J. Meynié	2352 m 172 m			
2008	X. Meniscus	−178 m	Endless shaft		
2012	X. Meniscus	−223 m	Endless shaft		
2016	X. Meniscus	−233 m	Horizontal gallery		Junction area

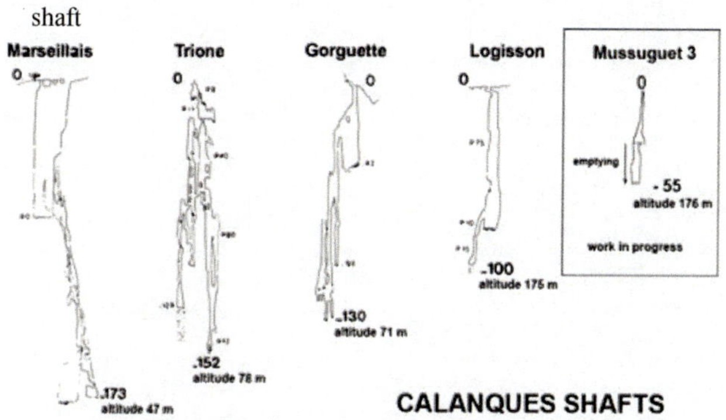

Fig. 7.1 Main chasms explored upstream of Port Miou and Bestouan in the Calanques massif (from Courbon and Parein 1991)

- Aven du Trione. This shaft, located near Cassis, reaches the altitude of 78 m a.s.l. and has a lake at its base. This cavity is unfortunately very dangerous because of the presence of CO_2.
- Gouffre du Logisson; This shaft is near the military camp of Carpiagne. Exploration did not go beyond the altitude of 175 m a.s.l.
- Gouffre du Mussuguet-3. This shaft is close to the Volanthen chamber in Le Bestouan cave. Cavers from the "Rivières Mystérieuses" group are currently emptying the clay and rock filling to find a passage.

In addition to Mussuguet-3, the chasms of Trione and Gorguette are particularly well located, and they show traces of water loading. Further exploration could make it possible to reach the galleries of Le Bestouan cave (see 8.3);

The large cave network of Petit-Saint-Cassien, a tenth of kilometers away from Cassis, is probably connected to the Port Miou system. Indeed, deep groundwater circulation is known in that cave whose destination could not be discovered by dye test. No spring was encountered except for the Foux-de-Nans spring, a temporary outlet where the water only flows at flood time. Thus, a deep drainage of the karst aquifer toward Cassis springs is very likely (Fig. 7.2).

Fig. 7.2 Geological section of plan d'Aups with Petit-Saint-Cassien chasm. (Cavalera 2007)

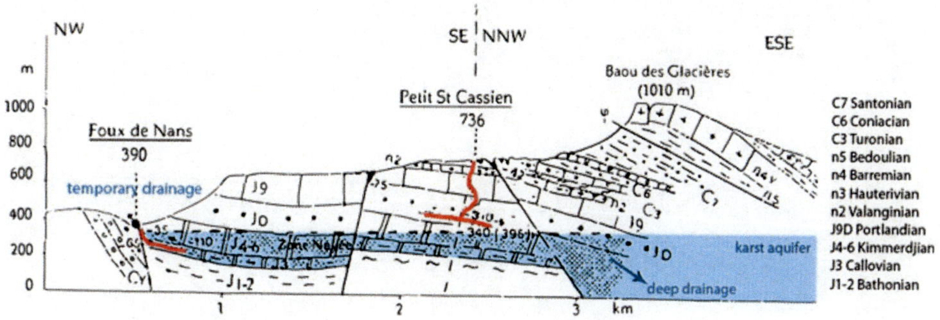

References

Acquaviva, G.: Resurgence marine de Port-Miou. ANAR Bull' **32**, 6–7 (2012)

Cavalera T.: Etude du fonctionnement et du bassin d'alimentation de la source sous-marine de Port Miou (Cassis). Approche multicritère. Thesis geol. University of Provence, Marseille, p. 397, (2007)

Courbon, P., Parein, R.: Atlas souterrain de la Provence et des Alpes de Lumière, p. 253. Gap edit, La Ravoire (1991)

Guieu, G., Ricour, J., Rouire, J.: Découverte géologique de Marseille et de son décor montagneux, p. 216. BRGM édit, Orléans (1996)

Survey of Deep Flooded Caves

8

8.1 Generalities

Surveying a submarine cave is a challenge as most of the diving time is devoted to exploration and as the quantity of air in the tanks is limited. Usually, the survey only consists in measuring the length of the safety line, in taking a few directions with the compass and in recovering the depths recorded by the dive computer.

However, more accurate work can be done using a tape, a dive compass and a depth meter. Scheme can be drawn with a pencil on a dive slate. I successfully used polyester paper on which it is possible to write with a pencil and erase underwater. It is easy to make small polyester booklets. On this kind of water proof paper it is also possible to make laser photocopies of previous notes and schemes to control them during subsequent dives.

Rebreathers that allow long duration in the water, dive computers and electronic tablets offer today new alternatives. However, whatever the methods are, the precision is rarely sufficient and submarine beacons can be used to precise, from the surface, the position of significant sites like rooms, air bells or gallery junctions. For instance, it makes it possible the drilling of a well to catch the water (Gilli 2015).

In Port Miou, different methods were used to survey the cave and several beacons were placed. Recently, automatic systems that use accelerometers, electronic compass and acoustic doppler velocimeter were used that gave valuable results for the deeper most part of the cave.

8.2 Port Miou

8.2.1 Geophysic Location

Geophysics technics were used to estimate the position of the flooded gallery like electromagnetic method (Fig. 8.1), a classical inductive method and a second method where one electrode was placed in the submarine river (Cornet et al. 1963). Both methods gave similar indications, but the precision was not sufficient.

8.2.2 Traditional Survey

The first survey of Port Miou was executed in 1968 between the entrance and the air bell that was previously discovered by the divers, 530 m from the entrance. As a dam was planned in the air bell, the aim of that survey was mainly to get a very accurate location to dig the well and the gallery from the surface. Traditional survey technics by professional divers were used (Courbon 2013).

The location was confirmed from the surface by using a magnetic beacon installed in the air bell. There was only a difference of 3 meters between the point given by the survey and that given by the magnetic determination. This made it possible to successfully dig the well and the gallery to reach the bell.

During the construction of the dam, a second survey was realized upstream of the air bell on a distance of 400 m. The survey stations were placed on the walls, along the safety wire that was installed by the divers. The length was measured with tapes, the directions with compass and the depths with diver depth meters.

The survey operations continued for several years after the SRPM project was abandoned. As the precision of this kind of survey is around 5%, a beacon was installed in the Swiss's air bell which made it possible to detect precisely its position from the surface for recalibrating the survey.

8.2.3 Automatic Surveying

Surveying deep parts of the flooded caves is another challenge as the diver cannot spend a long time for doing it. Automatic surveying is long overdue.

In Port Miou, a first trial was done using the CobraTac navigation device. It was designed for military purpose, but it is now used for reconnaissance and mapping applications. This navigation and survey device integrates the data from a

E. Gilli, *Port Miou and Le Bestouan (Cassis, France)*, Cave and Karst Systems of the World,
https://doi.org/10.1007/978-3-030-50192-1_8

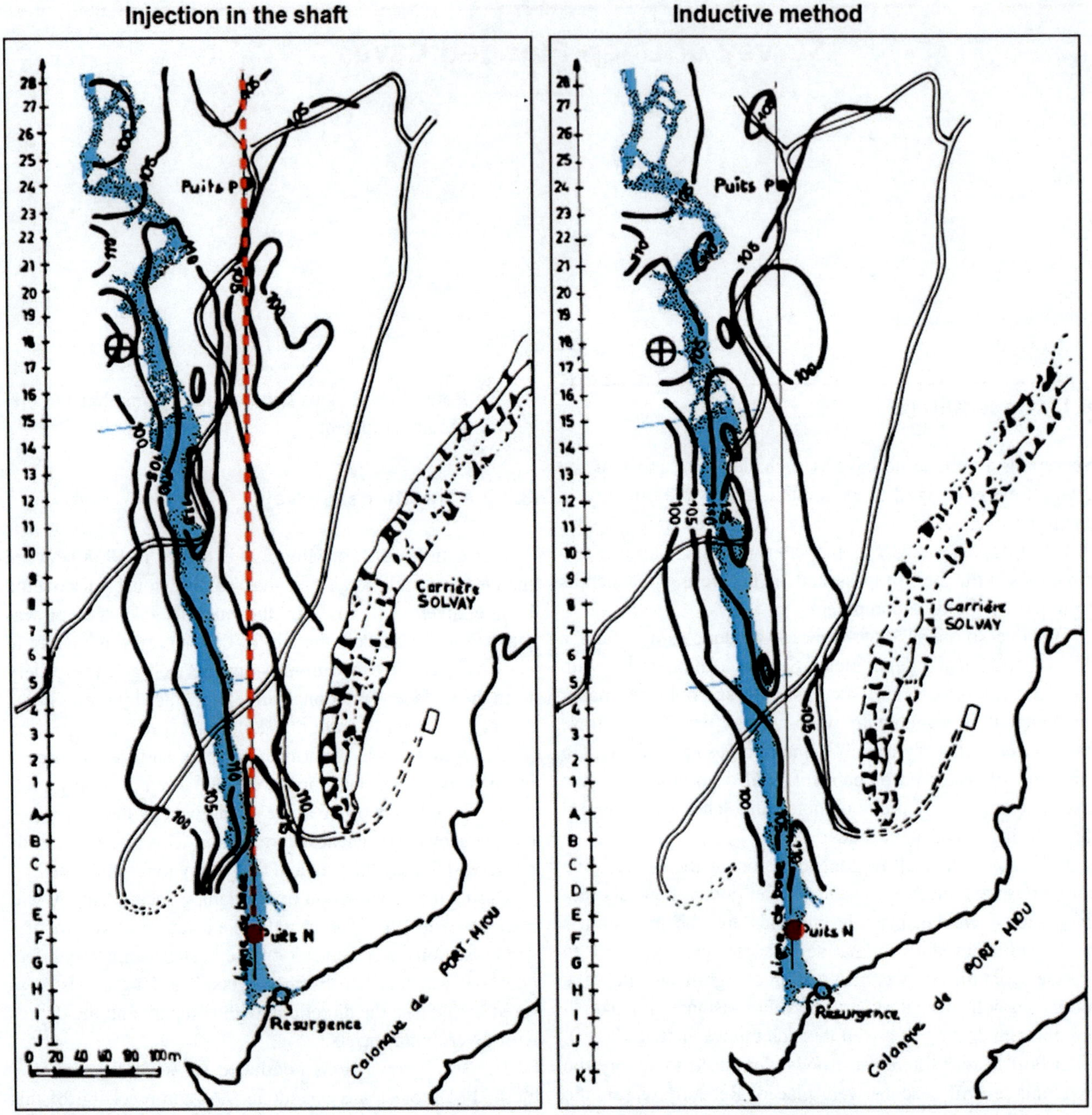

Fig. 8.1 Electromagnetic methods used to estimate the position of the gallery from the surface (from Cornet et al. 1963)

Doppler Velocity Log (DVL), fluxgate compass and pressure transducer, and this equipment allows a diver to quickly and easily survey the sea or lake bottom topography and make a georeferenced map. A test was done in Port Miou on an already mapped portion of the cave. The length was 400 m, and the survey was also made on the way back. Figure 8.2 presents a comparison between the classical survey and the CobraTac survey. One can note that the difference between the departure point and the arrival one is only 9 m.

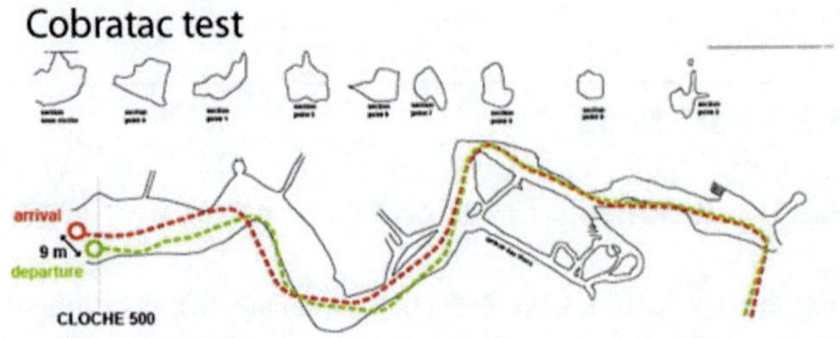

Fig. 8.2 Automatic survey upstream of the dam (from Courbon 2013)

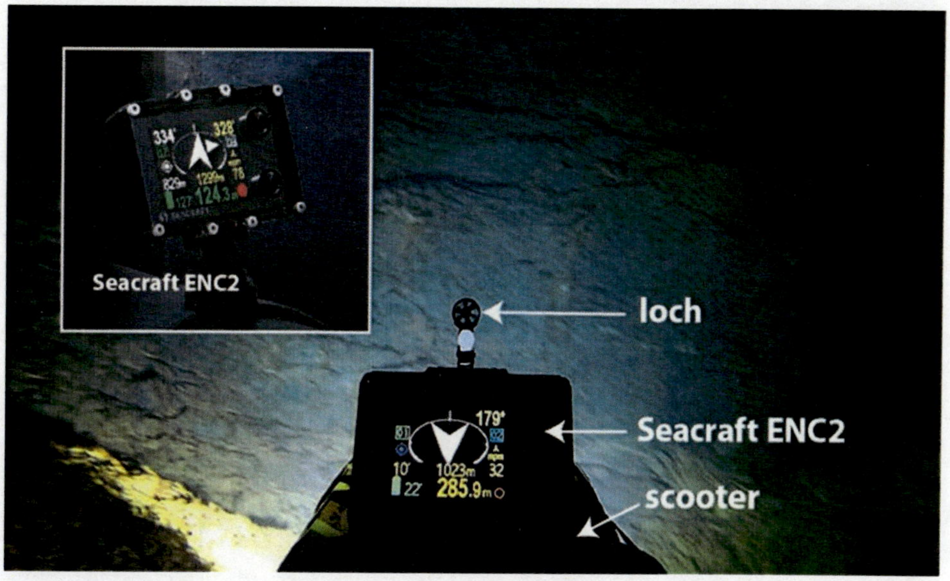

Fig. 8.3 Electronic navigation console ENC2 SEACRAFT (*Photo* Meniscus)

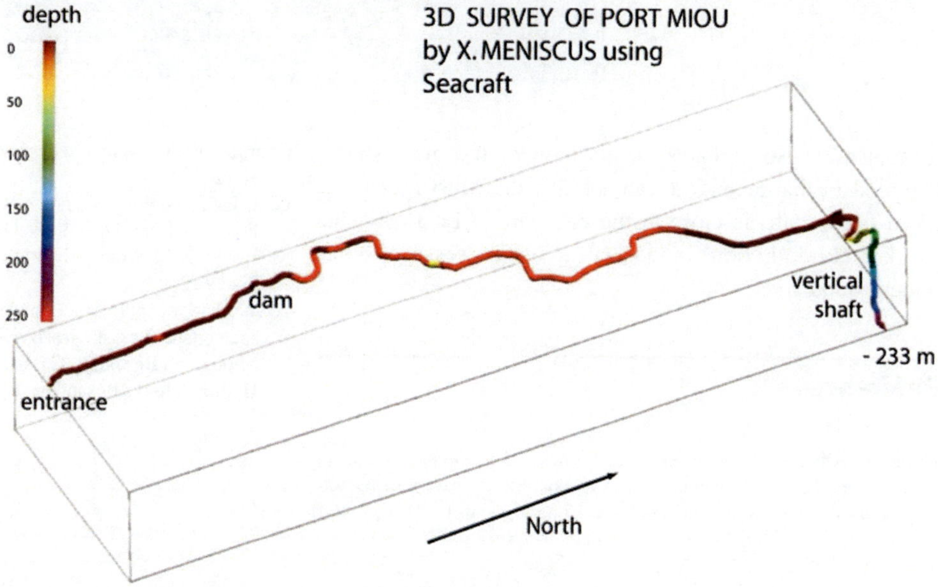

Fig. 8.4 Automatic survey of Port Miou by Meniscus using Seacraft ENC2 (from Meniscus 2017)

It is assumed that this error is mainly due to the equipment of the diver that creates a magnetic anomaly (Courbon 2013).

Another system, the Seacraft ENC2, an electronic navigation console, was tested by Meniscus (2017). He founds good concordance with the already existing surveys and now uses it for his deep explorations. The end part of Port Miou cave was surveyed with this equipment (Fig. 8.4) that is installed on the dive scooter. He used it in 2019 for his depth world record in Font Estramar where he reached −286 m (Fig. 8.3).

8.3 Le Bestouan

The survey of Le Bestouan cave was only made for the main gallery until the junction between the Flou gallery and the main gallery.

Both parts contain air bells; thus, beacons were placed to detect their position from the surface (Fig. 8.5). The following map presents Port Miou and Le Bestouan beacons

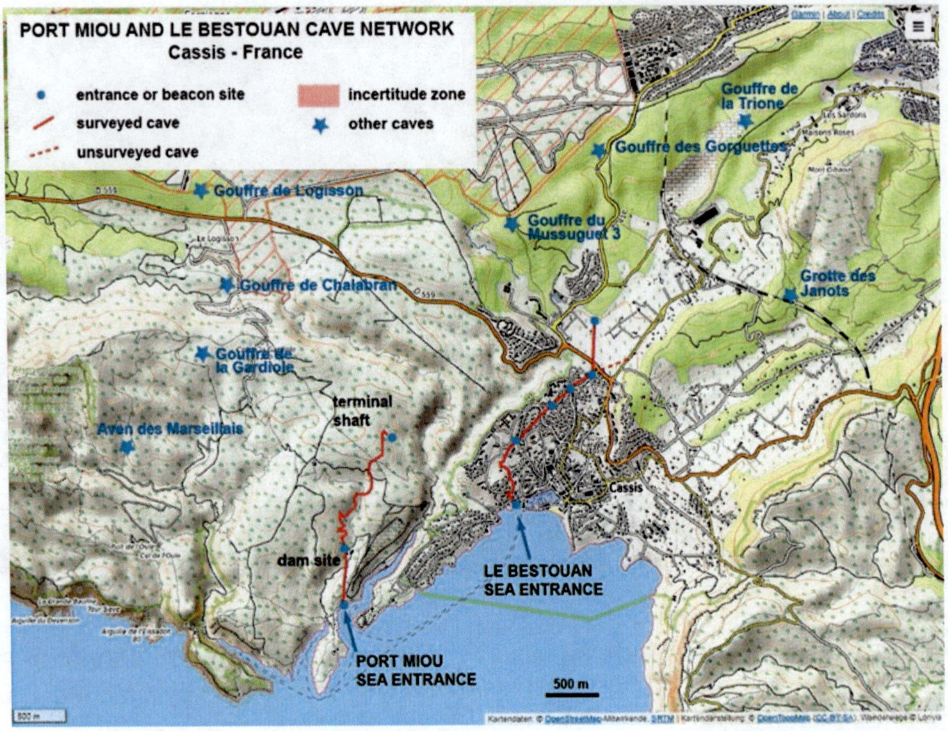

Fig. 8.5 Location of the underground beacons (from Arfib and Douchet 2011)

locations. It also contains the position of the main shafts explored in that area. One can see that the Mussuguet-3, a 55 m deep shaft, is close to the end part of Le Bestouan. Work is currently done in the shaft to open a passage to the underground river.

References

Cornet, G., Durozoy, G., Gouvernet, C., Munck F.: La source sous-marine de Port Miou (Calanques de Cassis). Etude par prospection géophysique. rapp. BRGM DS.63.A107. Orléans. p. 13, (1963)

Courbon, P.: Topographie des rivières noyées de Cassis. XYZ **137**, 33–38 (2013)

Gilli, E.: Karstology. Karsts, caves and springs. CRC Press, Taylor & Francis Group, New York USA, p. 244, (2015)

Meniscus, X.: Encore plus profond à Port Miou: −233. Spelunca **146**, 27–29 (2017)

Arfib, B., Douchet, M.: Etat des connaissances hydrogéologiques et spéléonautiques sur les rivières souterraines sous-marines de Port Miou et du Bestouan (Cassis, France). 9th Conference on Limestone Hydrogeology, Besançon, France, 1–3 September (2011)

Watershed Definition

9

9.1 Watershed Definition

In addition to the exploration of both Cassis submarine caves, the definition of the watershed is also a challenge. Classical dye tests are difficult to realize because the discharge of the springs is important (5–7 m^3/s), the superficy of the theoretical basin is more than 1000 km^2 and the size of the water body is huge. Indeed, they implicate the use of a great quantity of tracers and a long monitoring of the springs. Moreover, introducing high quantity of tracer in the local aquifers would affect wells and springs used for water supply. It would also make it impossible to realize dye tests for other studies, in that area, for a long time.

9.2 Structure

A first approach to define the watershed is the analysis of the geological structure. As both springs are located in the limestone series of the Beausset unit, it was supposed that they were draining only this unit. Thus, the superficy was first evaluated to 150 km^2.

9.3 Hydrological Balance

Another possible approach is observing the quantity of water produced by the karst units north of Port Miou springs. The observation of a water deficit means that part of the groundwater reaches another karst aquifer. The significant deficits noted in the Sainte Baume massif by C. Coulier and then P. Martin (Coulier 1985; Martin 1991) made it possible to widen the spring basin toward the north, including the southern flanks of Sainte Baume up to an area of approximately 200 km^2 (Guieu et al. 1996).

With a same approach, I proposed to extend the basin to other units that were suffering important water deficit: Aurelien, Allauch, Etoile, Ragagnas and Sainte-Victoire massifs (Gilli 2001). I extended it later to Brignoles and Loube massifs and part of Beausset basin (Blavoux et al. 2004). Then I asked Cavalera (2007) to precise the basin, but his work took place during an exceptional period of drought. He then calculated a reduced basin. Anyway, these research works prove a deficit for the following limestone massifs: Allauch, Etoile, Aurélien, Sainte-Victoire, Sainte Baume and Loube which leads the precited authors to propose a basin that includes all or part of these units the total superficiy being estimated between 500 and 1000 km^2 (Fig. 9.1).

Of course, the size of the basin depends on the quantity of water that emerges from Cassis springs. It was important to be sure that all the outlets were taken into account. A salinity study took place in 2009 to check if springs were present along the limestone coast, in addition to those of Port Miou and Le Bestouan (Gilli et al. 2009; Fournillon 2012). The study proved that except for a small spring in the Sugiton creek, that drains a local karst aquifer, no other springs are encountered in the Calanques massif (Fig. 9.2).

9.4 Dye Tests

Several dye tests were performed to try to explain the origin of Cassis springs water. Unfortunately, due to the large size of the watershed, only few of them were successful and they only gave information on the end part of the karst system (Fig. 9.3)

(1) *Coulin sinkhole* (Aubagne). A positive dye test was done in 1965 with 50 kg uranine in a sinkhole located in the southern part of Aubagne plain. The tracer was detected 42 days after the injection (Durozoy and Paloc 1966).

E. Gilli, *Port Miou and Le Bestouan (Cassis, France)*, Cave and Karst Systems of the World,
https://doi.org/10.1007/978-3-030-50192-1_9

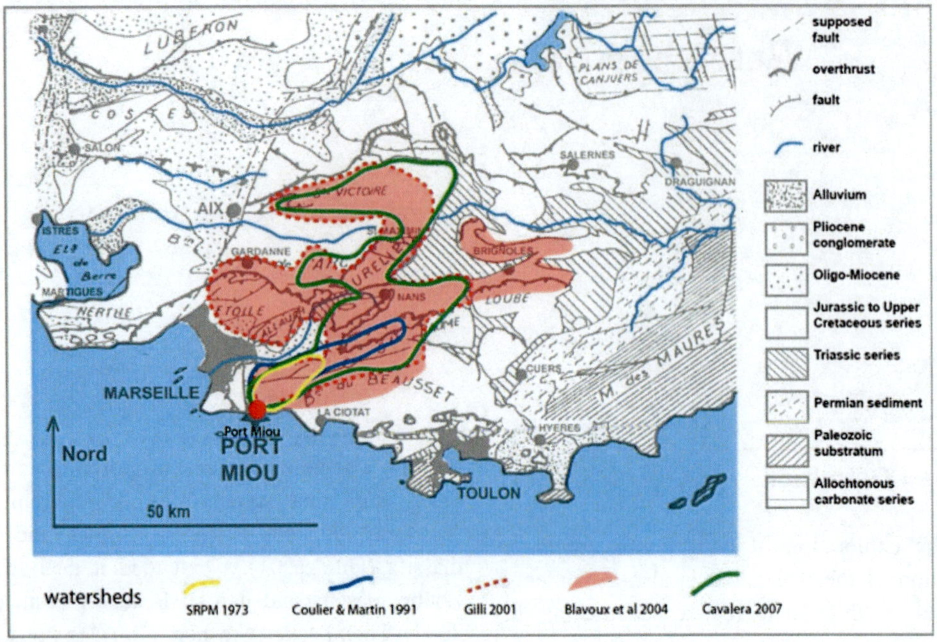

Fig. 9.1 Evolution of the watershed estimation

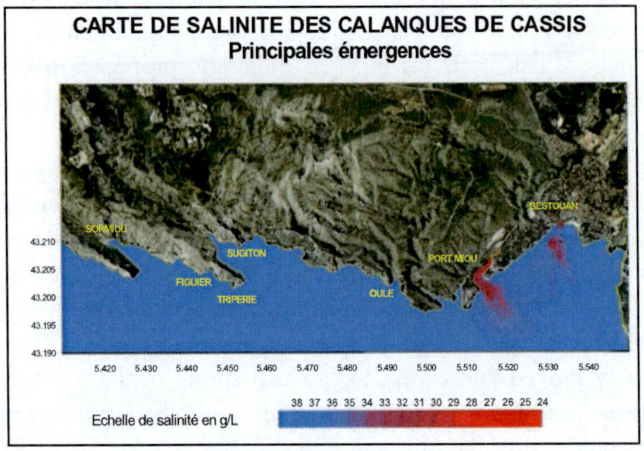

Fig. 9.2 Salinity map of Calanques massif. Except for Cassis, no important spring is observed (Gilli et al. 2009)

(2) *Mussuguet tunnel* (Cassis). A positive dye test was done in 1966, between both springs and an open crack discovered during the digging of Mussuguet train gallery (Durozoy and Paloc 1966). The speed was 32 m/h.

(3) *Mauregard sinkhole* (Ceyreste) (Arfib and Lamarque 2012). Sulforhodamine B was poured in a sinkhole during a rainy period. The positive transfer to Port Miou and Le Bestouan took only 7 days for a distance of 12 km. This dye test confirms the extension of the basin to the inner parts of Beausset unit proposed by Blavoux et al. (2004).

(4) *Chibron polje* (Signes). A dye test in Chibron polje was done in the 1970s by Durand (oral communication) who used 15 kg uranine. However, the detection of the tracer in Port Miou, six month after, was signaled but never confirmed.

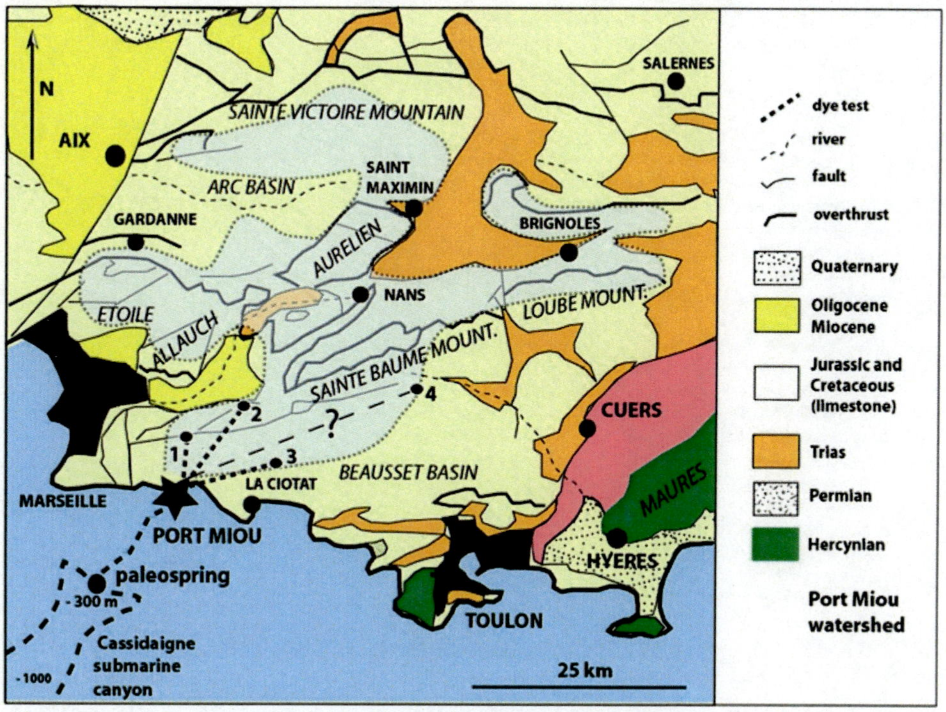

Fig. 9.3 Largest theoretical watershed and dye tests

References

Arfib, B., Lamarque, T.: Résultats préliminaires et premières interprétations du traçage KarstEAU du 08 février 2011 par injection de Sulforhodamine B à la perte de Mauregard (Ceyreste) (version du 09/02/2012), p. 14. Bestouan et Port Miou. Aix-Marseille Université, Résultats positifs sur les sources sous-marines de la baie de Cassis (2012)

Blavoux, B., Gilli, E., Rousset, C.: Alimentation et origine de la salinité de l'émergence karstique sous-marine de Port Miou (Marseille - Cassis – Bouches du Rhône). C. R. Geosciences, Elsevier, Paris **336**, 523–533 (2004)

Cavalera, T.: Etude du fonctionnement et du bassin d'alimentation de la source sous-marine de Port Miou (Cassis). Approche multicritère, p. 397. Thesis geol. University of Provence, Marseille (2007)

Coulier, C.: Hydrogéologie karstique de la Sainte Baume occidentale (Bouches-du-Rhône-Var, France). Thesis Geol. Aix-Marseille 1, p. 395, (1985)

Durozoy, G., Paloc, H.: Bassin du Beausset. Enseignements apportés par les expériences de coloration, BRGM Orléans (DS 69 SGL 212 PRC), (1966)

Fournillon, A.: Modélisation géologique 3D et hydrodynamique appliquées aux réservoirs carbonatés karstiques: caractérisation des ressources en eau souterraine de l'Unité du Beausset (SE France). Doct. thesis hydrogeol. AixMarseille University, p. 425 (2012)

Gilli, E.: Compilation d'anciennes mesures de débit à Port Miou. Apport à l'hydrogéologie de la Provence. 7e coll. hydrogeol. en pays calcaire et milieu fissuré. Besançon, 20–22 Sept 2001. pp. 157–160, (2001)

Gilli E., Fournillon A., Tassy A., Arfib, B.: Localisation des émergences karstiques des calanques et de la baie de La Ciotat. Projet Karsteau – Université de Provence, p. 49, (2009)

Guieu, G., Ricour, J., Rouire, J.: Découverte géologique de Marseille et de son décor montagneux, p. 216. BRGM édit, Orléans (1996)

Martin, P.: Hydromorphologie des géosystèmes karstiques des versants nord et ouest de la Sainte Baume (Bouches. du Rhône, Var; France). Etude hydrologique, hydrochimique et de la vulnérabilite à la pollution. Thesis Univ Provence. Aix-Marseille, (1991)

10 Developments and Conclusions

10.1 Current Research and Future Developments

10.1.1 Research

Research is currently going on in several directions. Cave divers and Riviéres Mystérieuse cavers group in Cassis try to go further in the exploration of both underground rivers but also on the different caves in that area that could be connected to the rivers. Marseille University tries to acquire a better knowledge of Port Miou river behavior and to define its watershed. The mechanism of salt contamination has also to be precised.

In addition, an important question remains unsolved. What is the relation between Port Miou and le Bestouan? They drain the same aquifer as proved by all dye tests that were positive for both springs. However, the relation between Port Miou and Le Bestouan remains unclear. Current explorations indicate that both caves converge, but the water quality is different and the rivers have a different flow regime. Le Bestouan river could be a recent leak of Port Miou aquifer that is now hollowing a new cave to adapt the aquifer drainage to the current sea level.

A better knowledge of the full system is of course fundamental to develop this water resource in a sustainable way. The local needs are important: irrigation, water supply and fire-fighting; we can then hope that the above actions will be facilitated by the community.

10.1.2 Site Valuation Projects

The first attempts to catch the water was by the SEM (Marseille water society) during the 1970s, but the failure of the underground dam project stopped the research until our new studies in the 2000s that explained the mechanism and the place of the marine intrusion.

However, since that date, there was no recent attempt for catching the resource in the basin, upstream of the contamination zone. The only places, in Cassis spring watershed, are the already existing Coulin water wells (depth 135 m, water table altitude 65 m), in Gemenos, and the Puyricard water well (depth 200 m, water table altitude 65 m), in Cuges-les-Pins. Both wells probably catch Port Miou aquifer, and the water they collect contains no chloride. New deep water wells could be planned upstream of the shaft.

A few projects are currently being examinated to reach both Port Miou and Le Bestouan rivers with drillings in view to catch the water. But as water remains brackish in Port Miou, this resource would be used to combat forest fires.

10.1.3 Geopolitics of Water and Water Lobbies

The local population is around 2 millions thus water supply is an important source of income. The water companies have a huge turnover, and there is a competition between surface water and groundwater to feed the population. Nowadays, the surface solutions, with long and costly aqueducts, are privileged (Fig. 10.1). They are highly supported by powerful water lobbies. So when I started the new studies of Port Miou, I was not surprised to receive a letter from the Head of Societé-des-Eaux-de-Marseille asking me to immediately stop any research on Port Miou site. We were starting Cavalera's thesis and, of course, I refused to stop the work. Then, to save the research project, my student had to make all the fieldwork at night time or during his holidays. The project of digging a gallery, the Janots Gallery, for installing a new pipeline to supply Cassis and La Ciotat area, was probably the real cause of this demand (Fig. 10.1). The tunnel was achieved in 2018 (Willis 2019). The pipeline discharge will be up to 1 m^3/s.

E. Gilli, *Port Miou and Le Bestouan (Cassis, France)*, Cave and Karst Systems of the World,
https://doi.org/10.1007/978-3-030-50192-1_10

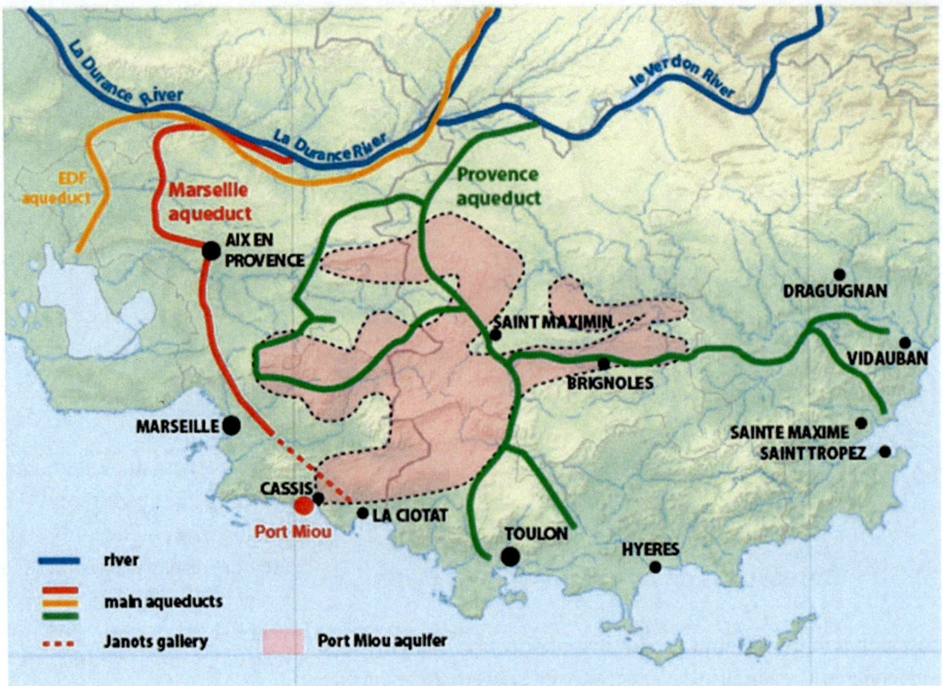

Fig. 10.1 Aqueducts in the area of Port Miou basin

10.2 General Conclusion

Port Miou and Le Bestouan adventure is far to be achieved. However, we know more about the Moon or Mars than on both caves although they are located in the perimeter of Marseille, the third city of France. More people walked on the Moon that in the end part of Port Miou.

Going further in the knowledge of this huge karst aquifer is a real challenge and new methods and techniques are necessary. However, human exploration has its limits, and the commitment quickly becomes irrelevant. In Port Miou, the current record is 233 m deep, more than 2 km from the entrance. Like for planets, the future of Port Miou exploration will probably involve manned or unmanned underwater robotic probes. The Remote Operated Vehicles (ROVs) take over from the man and allowed to explore the Fontaine de Vaucluse up to 308 m deep or the Pozzo del Mero up to 320 m. However, if a ROV can easily work in open water when equipped with a zero-floating link cable, horizontal movement over a long distance is impossible. Current ROVs can therefore only explore pseudovertical wells. For horizontal networks, there is currently no device other than that used for the exploration of the Titanic (Cameron Ltd.). It is driven by an optical fiber that unfolds as it progresses and abandons on the spot once the exploration is complete. A slowly water soluble material is used to allow the total degradation of the fiber in a few months.

Cave exploration is the main challenge, but it is not the only one. Indeed being able to know precisely from where the water comes is very difficult. Due to the important size of the watershed and the imprecision on the geological and structural settings, it is difficult to estimate the relationships between the different karst units of Basse Provence.

Dye tests could help to define the bassin but due to the distances, the quantity of products (uranine, rhodamine) is very important. The monitoring of the springs would last several months and, after one experience, the presence of remaining dye in the aquifers for several months, or years, would be problematic for other dye tests.

Looking for anthropogenic products locally produced by wastes or industries could be a solution. The local industries are developed since the twelfth century (e.g., Marseille Soap) until now. For a long time, pollution standards were totally ignored and different chemical products were stored without any protection, or poured in wastes. The leaching of these products by rainwater provokes their migration to the groundwater. Our hypothesis was as follows: If different anthropogenic products are present in a specific site, like for instance an industrial waste, they could form of a local signature. This anthropogenic pollution could be an opportunity to define more precisely the basin, subject to be able to detect very low concentrations. The red mud poured by the alumina factory that was used efficiently (see chapter…) perfectly illustrates this theory. A similar approach could be used elsewhere, subject to identify several sources of pollution in diverse areas of the theoretical impluvium.

With Cavalera (2007), we tried to study thunders isolated on a part of the theoretical impluvium and see if there was pressure effects that could augment the discharge of the springs but unfortunately the studied period was a very dry one and our project failed. This trial could be engaged again as it was successfully used for a similar study in the area of Nice (Southeastern France) (Gilli 1999). The data collected by Marseille University since several years, coupled with meteorological data, would facilitate such a study.

But whatever are the solutions Cassis springs' adventure will continue to be a story of human exploration, technology, science and socioeconomy. These springs are a mystery since the Antiquity, and we shall probably never be able to know the precise origin of the water, but as Mark Twain said: "They didn't know it was impossible so they did it."

Let's do it…

References

Cavalera, T.: Etude du fonctionnement et du bassin d'alimentation de la source sous-marine de Port Miou (Cassis). Approche multicritère. Thesis geol., p. 397. University of Provence, Marseille (2007)

Gilli, E.: Détection de sources sous-marines et précision de l'impluvium par mesure des variations de salinité. L'exemple de la source de Cabbé-Massolins (Roquebrune-Cap-Martin, France) CRAS Paris, **IIa-329**, 109–116, (1999)

Willis, D.: Overcoming Multiple Caverns: Successful TBM Tunneling in Karstic Geology at Galerie des Janots. Tunnel. https://www.tunnel-online.info/en/artikel/tunnel_Overcoming_Multiple_Caverns_Successful_TBM_Tunneling_in_Karstic_Geology_3413869.html (2019). Accessed on 9 Jan 2020

Videos

Port Miou

The cave divers Nicolas Andreini, Xavier Meniscus, Frédéric Swierczynski and Michael Walz provided unestimable video shots of Le Bestouan and Port Miou, including the deepest parts.

Meniscus X., 2016, *Plongée dans la résurgence de Port Miou : du barrage jusqu'à -145m*. https://vimeo.com/169161160

Meniscus X, 2019, Topographie ENC2 −225 m avril 2019. https://www.youtube.com/watch?v=NfLPXjFRbNY

Andreini N., Walz M., and Meniscus X, 2016, *Port Miou: Une plongée dans ses abyss jusqu'à −203 m*. https://vimeo.com/185379302

Swierczynski F., 2016, Cave diving, Port Miou 2016 (U-W-X) https://vimeo.com/161296247

Walz M., 2016, Topographie dans le puits terminal de Port Miou. https://vimeo.com/185720869

Walz M., 2016, Port Miou Mai 2016. https://vimeo.com/169014677

Le Bestouan

Ludovic Dendeloeuf also provided some video shots of the first part of Le Bestouan.

Dendeloeuf L, 2019, Tuto de plongée n°6 «Le Bestouan à Cassis». https://www.youtube.com/watch?v=XBZRiJGa1Ho

E. Gilli, *Port Miou and Le Bestouan (Cassis, France)*, Cave and Karst Systems of the World,
https://doi.org/10.1007/978-3-030-50192-1